KENKEN®

A PUZZLE A DAY!

365 Puzzles That Make You Smarter

KenKen: Math & Logic Puzzles That Will Make You Smarter!
Series editor: Robert Fuhrer (*KenKen Puzzle, LLC, USA*)

ISSN: 2529-8003

Vol. 1: *The KENKEN Method — Puzzles for Beginners*
150 Puzzles and Solutions to Make You Smarter
ed. R Fuhrer

Vol. 2: *KENKEN: Lim-Ops, No-Ops and Twist!*
180 6 × 6 Puzzles That Make You Smarter
ed. R Fuhrer

Vol. 3: *KENKEN: A Puzzle A Day!*
365 Puzzles That Make You Smarter
ed. R Fuhrer

KenKen: Math & Logic Puzzles That Will Make You Smarter! **Volume 3**

KENKEN®

A PUZZLE A DAY!

365 Puzzles That Make You Smarter

Created by

Tetsuya Miyamoto

Edited collection by

Robert Fuhrer

Founder, KenKen Puzzle, LLC

NEW JERSEY • LONDON • SINGAPORE • BEIJING • SHANGHAI • HONG KONG • TAIPEI • CHENNAI • TOKYO

Published by

World Scientific Publishing Co. Pte. Ltd.

5 Toh Tuck Link, Singapore 596224

USA office: 27 Warren Street, Suite 401-402, Hackensack, NJ 07601

UK office: 57 Shelton Street, Covent Garden, London WC2H 9HE

Library of Congress Cataloging-in-Publication Data

Names: Miyamoto, Tetsuya, 1959– creator. | Fuhrer, Robert, editor.

Title: Kenken: a puzzle a day! : 365 puzzles that make you smarter / created by Tetsuya Miyamoto ; edited collection by Robert Fuhrer, founder, KenKen Puzzle, LLC.

Description: 1st edition. | New Jersey : World Scientific, 2020. | Series: Kenken: math & logic puzzles that will make you smarter!, 25298003 ; vol. 3

Identifiers: LCCN 2019052121 | ISBN 9789813236684 (hardcover) | ISBN 9789813235878 (paperback)

Subjects: LCSH: KenKen. | Logic puzzles. | Mathematical recreations.

Classification: LCC GV1493 .M5227 2020 | DDC 793.73--dc23

LC record available at https://lccn.loc.gov/2019052121

British Library Cataloguing-in-Publication Data

A catalogue record for this book is available from the British Library.

For any available supplementary material, please visit
https://www.worldscientific.com/worldscibooks/10.1142/10874#t=suppl

Printed in Singapore

CONTENTS

INTRODUCTION

Hello, puzzle enthusiast!

Congratulations on your exceptional decision. Picking up a copy of *KenKen A Puzzle A Day* may be one of the best choices you make all year.

KenKen puzzles were created by Tetsuya Miyamoto to help his students improve their calculating ability, logical thinking, problem solving, and patience. Solving KenKen puzzles will help you improve these skills, too, though your patience will sometimes be tested. KenKen puzzles provide a unique combination of fun and enrichment, and while doing them, you may become so engrossed that you actually forget you're doing arithmetic! Solving them will improve your computational skills, or at least reinforce them, and you may develop a deep interest in mathematics, seeing the elegance in numbers and their infinite combinations. At a minimum, KenKen will definitely provide a temporary distraction from the chaos of the outside world. You likely have seen traditional KenKen puzzles before — they appear in the *New York Times*, the *The Times* (UK), in myriad other international publications, on the KenKen app and website, and in several dozen books that you can find at your favorite retailers.

In this book, you'll find a KenKen puzzle for every day of the year. That's right — 365 puzzles, ranging in size from 4×4 to 9×9, and ranging in difficulty from easy to intermediate to advanced. We're hoping that you find time each day to indulge in just one intriguing puzzle.

What's more, you'll find three different types of KenKen puzzles, variants of the traditional ones, which use the typical rules of KenKen: lim-ops puzzles, which limit the puzzle to only using *some* of the four operations; "no-ops" puzzles, in which the operations are not indicated in the cages, only the target numbers; and twist puzzles, which require you to use a different set of numbers to fill in the grid.

Traditional KenKen puzzles require you to fill in an $n \times n$ grid with the numbers 1 through n. The grid is divided into cages indicated by the thick lines, and each cage contains a target number and an operation. The values to be placed in a cage must combine, in any order, to yield the target number using the given operation. For instance, the numbers in a [10×] cage must yield 10 when multiplied.

Lim-ops puzzles (or limited operations puzzles) are traditional KenKen puzzles but instead of using all four elementary arithmetic: +, plus (addition); −, minus (subtraction); ÷, obelus (division); ×, times (multiplication), you will experience different combinations of those. In this volume, the lim-ops puzzles are a mix of addition/subtraction only, and multiplication only.

To provide an additional challenge, there are also no-ops puzzles, in which no operations are provided in the grid! This may sound a little nuts, and it is, but it's also really fun. It means that there will be more possibilities to consider for every cage. In a 4×4 traditional puzzle, a three-cell [8×] cage would only have one possibility, {1, 2, 4}, while a [8+] cage would have three possibilities, {1, 3, 4}, {2, 2, 4}, and {2, 3, 3}. But in a 4×4 no-ops puzzle, a three-cell [8] cage could be filled with any of those four possibilities, because no operation has been indicated. These puzzles are not for the faint of heart. You'll need to consider lots of combinations to determine the correct values for each cage.

If that's still not enough for you, there are also KenKen Twist puzzles, which will really get your blood pumping. Instead of using 1 through 6 like you would in a 6×6 traditional puzzle, these diabolical creations require that you use a different candidate set for each puzzle. In one puzzle, you might use {1, 2, 3, 7, 8, 9}; in the next, {2, 4, 6, 7, 8, 9}. It's fun to use different numbers and consider new combinations. But stay on your toes! If you let down your guard, you may find yourself solving a puzzle with the wrong values!

Solutions can be found in the back of the book ... but no cheating! Try solving each puzzle before checking your work.

Tetsuya Miyamoto always says "The rival is not the person next to you. It's the YOU from yesterday. The more you challenge yourself, the more you will improve. Day by day, KenKen will help you grow!" To support Miyamoto-sensei's vision, there is a Solving Time box next to each puzzle so that you can track your progress and see how your skills improve. As you continue to solve KenKen puzzles, you will be able to increase your speed while maintaining accuracy.

Partaking in puzzles regularly has been shown to help people overcome innumeracy, delay or reduce memory loss, and enhance problem-solving skills. While the puzzles in this book get slightly harder from day to day, the progression in difficulty is designed to be gradual and logical. At the end of the year, when you look back at the first few puzzles in this book, you'll be amazed at the progress you've made and proud of how far you've come.

Every KenKen aficionado has a preferred setting and routine — with a cup of morning coffee, during a study break, just before bed. Whatever your preferred time or place, try to make KenKen part of your daily routine, because the more you practice, the better you'll get. And if you'd like even more KenKen in your life, visit our website at **www.kenkenpuzzle.com** for an endless supply of free puzzles!

Enjoy this book, and have a wonderful year! Happy solving!

THE RULES OF STANDARD KENKEN

Your goal is to fill in the whole grid with numbers, making sure no number is repeated in any row or column.

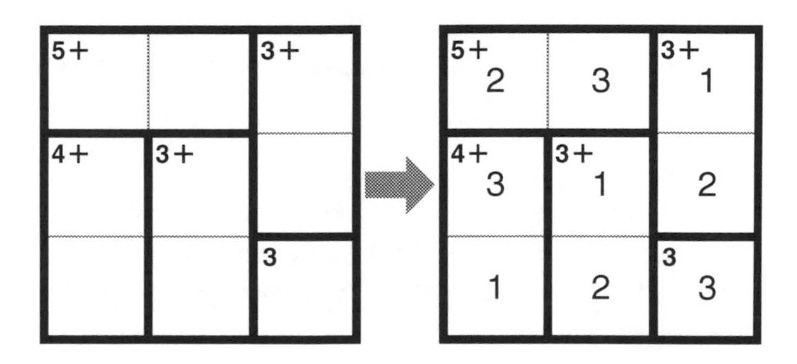

In a 3×3 puzzle, use the numbers 1–3.

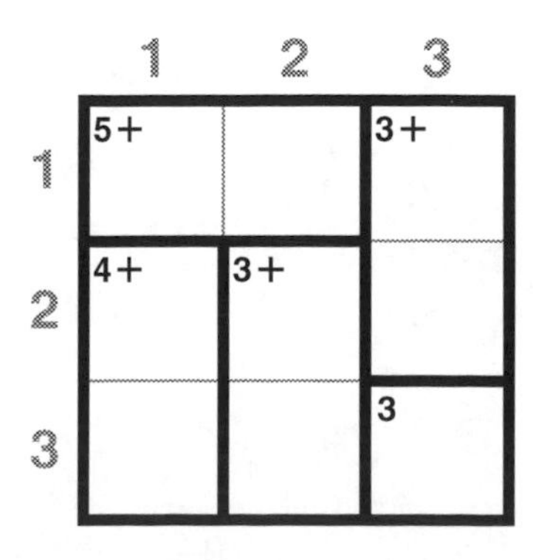

In a 4×4 puzzle, use the numbers 1–4.
In a 5×5, use the numbers 1–5, and so on.

5+		3+
4+	3+	
		3

The top left corner of each cage has a "target number" and math operation. The numbers you enter into a cage must combine (in any order) to produce the target number using the math operation noted (+, −, ×, or ÷).

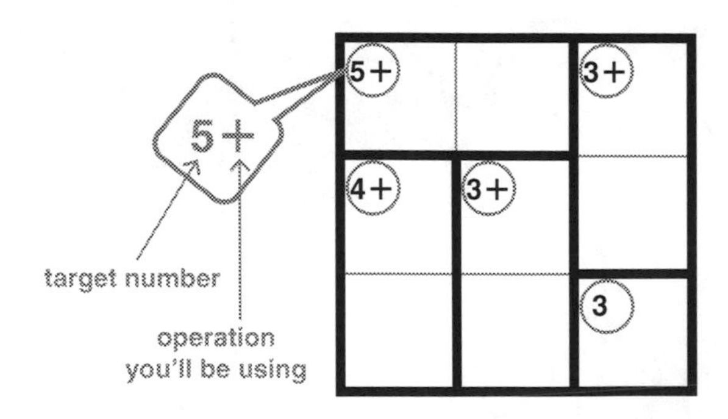

In this cage, the math operation to use is addition, and the numbers must add up to 5. Since the cage has 2 squares, the only possibilities are 2 and 3, in either order (2+3 or 3+2=5).

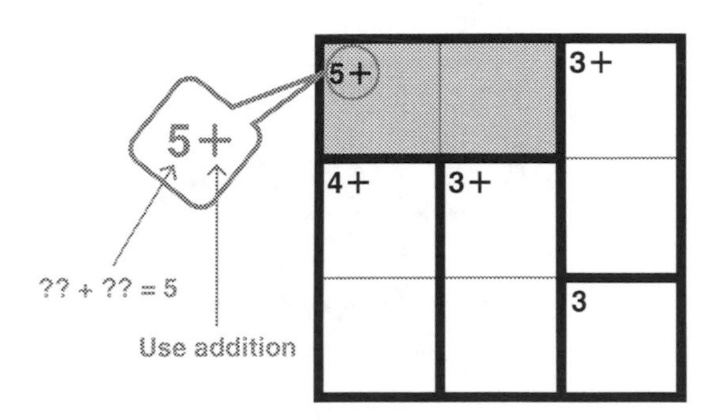

A cage with one square is a "Freebie"... just fill in the number you're given.

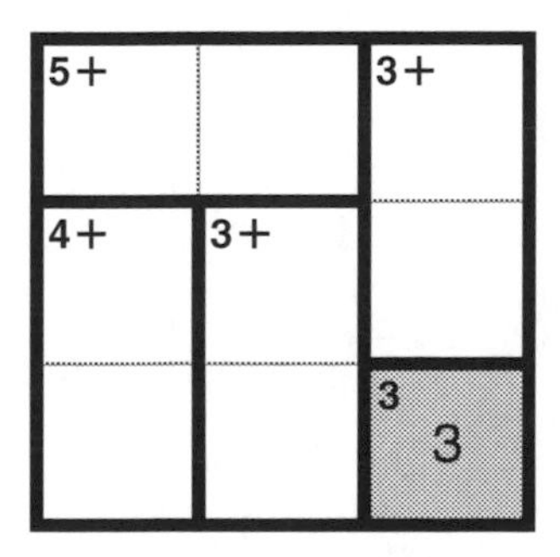

A number cannot be repeated within the same row or column.

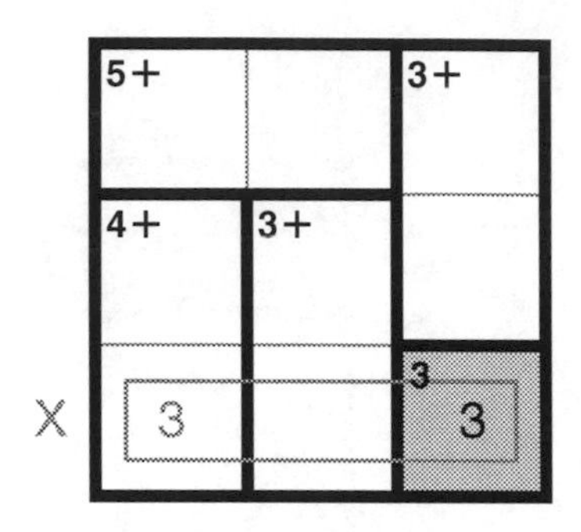

And that's it!

Solving a KenKen puzzle involves pure logic and mathematics.

No guesswork will ever be needed, and each puzzle has only one solution. So sharpen your pencil, sharpen your brain, and get started! In an instant, you'll know why Will Shortz, NPR Puzzle Master and The New York Times Puzzle editor calls KenKen "The most addictive puzzle since Sudoku!"

STEP-BY-STEP TUTORIAL FOR ALL PUZZLE TYPES

Lim-Ops Puzzles

Take it to the limit

In a typical KenKen puzzle, each cage contains a target number and an operation. In a lim-ops puzzle — short for "limited operations" — some of the operations won't be used. That is, the puzzle is limited to three, two, or even just one operation. In this volume, the lim-ops puzzles may contain only addition, only subtraction, a mix of addition and subtraction, or only multiplication.

Hints for lim-ops puzzles

To solve lim-ops puzzles, you'll follow the same process as solving standard KenKen puzzles, but the following hints may prove helpful:

- Subtraction cages with a target number that is one less than the grid size can only have a single set of numbers as the answer. For instance, a [5−] cage in a 6×6 puzzle must be filled with {1, 6}.

- Addition cages with a target number that is one less than twice the grid size can only have a single set of numbers as the answer. For instance, a [11+] cage in a 6×6 puzzle must be filled with {5, 6}.

- Because every value is used exactly once in a row or column, the sum of the values is the same for every row and column in a puzzle. For a 6×6 puzzle, the sum of the values is always $1+2+3+4+5+6=21$. This can be hugely helpful. If a row contains, say, a two-cell [10+] cage and a three-cell [8+] cage, the sum of those two cages is $10+8=18$; since $21-18=3$, the remaining cell must be filled with a 3. Or if there is a two-cell [4−] cage in a column for which the other four cells have a sum of 13, then the cage must be filled with {2, 6}, because $21-13=8$ and $2+6=8$.

- Consider whether a value must be odd or even (mathematicians refer to this as the *parity* of a number). For instance, let's say a row in a 6×6 puzzle contains a three-cell [11+] cage and a two-cell [3–] cage. The possibilities for the [3–] cage are {1, 4}, {2, 5}, and {3, 6}. In every case, the sum of the possible values for the [3–] cage are odd, so the sum of the values in the [11+] cage and the [3–] cage must be even. Since the sum of the values in a row in a 6×6 puzzle is always 21 (odd), the remaining cell in that row must be odd. While this may not give you the value for the cell directly, it reduces the number of possibilities by half.

A lim-ops example

The lim-ops puzzle below contains both addition and subtraction.

6+	17+		9+		
			2–	9+	
3–	5+			8+	
	9+	9+			5–
10+		8+	4–		
			10+		

This puzzle has one "gimme." The [17+] cage must be filled with {5, 6, 6}, as follows:

6+	17+ 6	5	9+		
		6	2–	9+	
3–	5+			8+	
	9+	9+			5–
10+		8+	4–		
			10+		

Every row and column of a 6×6 puzzle uses each value 1 through 6 exactly once. Consequently, the sum of every row and every column in a 6×6 puzzle is $1+2+3+4+5+6=21$. This information is very helpful for all types of KenKen puzzles, but it is especially useful with lim-ops puzzles.

- Two cells in the top row of the puzzle contain {5, 6} in the [17+] cage, and there is a [9+] cage at the end of the row. The sum of those five cells is $5+6+9=20$, so the first cell in the row must be $21-20=1$.

- The two cells in the second row of the [6+] cage must contain {2, 3}, since a 1 was used in the top cell of that cage. Further, there is already a 6 in the third cell of the second row, and there's a [9+] cage at the end of the row. The sum of those five numbers is $2+3+6+9=20$, so the top cell of the [2–] cage must be $21-20=1$. And if the top cell in the [2–] cage is a 1, then the bottom cell of that cage must be a 3.

- The third row contains a [5+] cage and a [8+] cage, and the bottom cell of the [2–] cage is a 3. The sum of those five cells is $5+3+8=16$, so the first cell in the third row must be $21-16=5$. And if the top cell of the [3–] cage is a 5, then the bottom cell must be a 2.

- Since the [3–] cage contains a 2, the 2 in the [6+] cage must occur in the second column, and the 3 is then in the first column.

With just this little bit of work, the puzzle is more than a quarter complete:

6+ 1	17+ 6	5	9+		
3	2	6	2– 1	9+	
3– 5	5+		3	8+	
2	9+	9+			5–
10+		8+	4–		
			10+		

As is typical with lim-ops puzzles, information from two (or more!) cages now needs to be used in tandem to make any progress.

- The [5+] cage cannot be filled with {2, 3}, because there is already a 3 in the third row. Therefore, the [5+] cage must be filled with {1, 4}. That means that the [8+] cage must be filled with {2, 6}. Since the [5–] cage in the last column must be filled with {1, 6}, the 6 in the [8+] cage must be in the left cell of that cage.

- The fourth row contains a 2 and a [9+] cage. Together, their sum is $2+9=11$, which means the other two cells must have a sum of 10. If the top cell of the [5–] cage were a 1, then the top cell of the [9+] cage would have to be a 9, which is impossible. Therefore, the 6 is in the top cell of the [5–] cage, and the top cell of the [9+] cage must be a 4.

Don't look now, but 16 of the 36 cells have been filled. Are we having fun yet? You betcha!

6+ 1	17+ 6	5	9+		
3	2	6	2– 1	9+	
3– 5	5+		3	8+ 6	2
2	9+ 4	9+			5– 6
10+	5	8+	4–		1
			10+		

With nearly half of the puzzle complete, things start to move quickly.

- The [5+] cage must be filled with {1, 4}, and since there is already a 4 in the second column, the order is dictated.
- The lower left cell of the three-cell [8+] cage must be 3, as it is now the only value not used in the second column.
- The two unfilled cells in the [8+] cage can either be {1, 4} or {2, 3}, but since a 4 already occurs in the third column, then it must be {2, 3}. Their order is determined because of the 3 in the lower left corner of the cage.

- The remaining unfilled cell in the third column (as part of the [9+] cage) must be a 1, as it is the only value not yet used.

- The other two cells in the [9+] cage must be {3, 5}, and the 3 in the [2–] cage above dictates their order.

The puzzle is now two-thirds complete.

6+ 1	17+ 6	5	9+		
3	2	6	2– 1	9+	
3– 5	5+ 1	4	3	8+ 6	2
2	9+ 4	9+ 1	5	3	5– 6
10+	5	8+ 3	4–		1
	3	2	10+		

One last piece of logic is needed to crest the hill.

- The bottom row contains {2, 3} in the [8+] cage, and there is also a [10+] cage. The sum of these five numbers is $2+3+10=15$, so the first cell in the bottom row must be $21-15=6$. The other number in the two-cell [10+] cage must be 4.

The hard work is now done, and the remainder of the puzzle becomes a formality.

- The only numbers missing from the fifth row are {2, 6}, and the 6 used in the [8+] cage in the third row dictates the order of {2, 6} in the [4–] cage.

- The three-cell [10+] cage in the bottom row must be filled with {1, 4, 5}, and the 1's and 5 in the cages above are enough to dictate their order.
- The two-cell [9+] cage in the second row must be filled with {4, 5}, and the three-cell [9+] cage in the first row must be filled with {2, 3, 4}, and their order is dictated by the other numbers already in those columns.

And there you have it, *c'est fini*.

6+ 1	17+ 6	5	9+ 2	4	3
3	2	6	2– 1	9+ 5	4
3– 5	5+ 1	4	3	8+ 6	2
2	9+ 4	9+ 1	5	3	5– 6
10+ 4	5	8+ 3	4– 6	2	1
6	3	2	10+ 4	1	5

As this puzzle shows, lim-ops puzzles can be deceptive. With only two operations, they appear to be less complex than other KenKens, but don't let their simplicity fool you. Addition and subtraction cages often have more possible combinations than multiplication or division cages, so more information is needed to eliminate the incorrect candidates.

No pain, no gain. Lim-ops puzzles may take you longer than other types of KenKens, but you'll also feel a great sense of satisfaction (and relief!) when you complete one, because of the effort that is required.

No-Ops Puzzles

Cancel the surgery ... no operations!

In a traditional KenKen puzzle, each cage contains a target number and an operation. The values to be placed in that cage must combine, in any order, to yield the target number using the given operation. For instance, the numbers in a [10×] cage must yield 10 when multiplied together.

With no-ops puzzles, each cage contains a target number but *no operation*. It means that *any* operation can be used to yield the target number. For instance, in the no-ops puzzle below, the numbers that fill the [10] cage must yield 10 when addition, subtraction, multiplication, or division is applied. The challenge is, you don't know which operation to use, so most cages in a no-ops puzzle will have far more possibilities than the cages in a standard KenKen puzzle. If it were a [10+] cage, the values would have to be {4, 6}. If it were a [10×] cage, the values would have to be {2, 5}. But since it's a [10] cage in a no-ops puzzle, it could be filled with either combination.

6		17		240	
	2			11	
8	11				
	90			12	
15			4		6
		10			

Hints for no-ops puzzles

To solve no-ops puzzles, you'll follow the same process as solving standard KenKen puzzles, but the following hints may prove helpful:

- Look for big numbers. In a 6×6 no-ops puzzle, a two-cell cage with a target number greater than 12, a three-cell cage with a target number greater than 18, and, in general, an *n*-cell cage with a target number greater than 6*n* means that the operation must be multiplication.
- Look for prime numbers larger than the grid size. For instance, in the puzzle above, the [11] and [17] cages imply addition, because it's not possible to get a product of either 11 or 17 with multiplication. (If the prime number is less than the grid size, however, such as 2, 3, or 5 in the case of a 6×6 puzzle, every operation is a possibility.)

A no-ops example

It's not clear where to start with the 6×6 no-ops puzzle below. That's often the case; it may take some exploration of what you know before you'll be able to fill in any cell.

6		17		240	
	2			11	
8	11				
	90			12	
15			4		6
		10			

Most of the cages in the puzzle above have multiple possibilities. As already mentioned above, the [10] cage has two possibilities: {4, 6} with addition or {2, 5} with multiplication. The [15] cage is a bit more challenging, because it could be either addition or multiplication, and values can be repeated in the cage because of its L-shape. Consequently, there are three possibilities for the [15] cage: {1, 3, 5}, {4, 5, 6}, {3, 6, 6}.

But the [17] cage at the top must be addition — no three values can multiply to 17 — and the only possibility is {5, 6, 6}. Because it is an L-shaped cage, the placement of each value is uniquely determined. This is kind of a big deal.

- The [17] cage requires two 6's, so the [10] cage in the bottom row cannot contain a 6. Consequently, it must be filled with {2, 5}, and because there is a 5 in the [17] cage, the placement of the 5 in the bottom row is determined.
- The [11] cage in the fifth column must be {5, 6}, and since there is already a 6 in the second row, the order of the values is determined.

Together, these insights form a very good start.

6		17 6	5	240	
	2		6	11 5	
8	11			6	
	90			12	
15			4		6
		10 5	2		

The [240] cage must be multiplication, because no other operation could generate a target number of that magnitude.

- The possibilities are {2, 4, 5, 6} or {3, 4, 4, 5}, but a 6 has already been placed in each of the first three rows. Hence, it must be {3, 4, 4, 5}. Because the 5 must be used in the third row, and because the set contains two 4's, the placement of all four values is uniquely determined.

- The [12] cage can either be {3, 4} or {2, 6}, but since a 4 has now been used in both the fifth and sixth columns, it must be {2, 6}, and their order is dictated by the values above.

We're getting somewhere! In fact, we're more than one-third done already.

6		17 6	5	240 4	3
	2		6	11 5	4
8	11			6	5
	90			12 2	6
15			4		6
		10 5	2		

Things are really starting to fall into place!

- The [8] cage can be either {2, 6} with addition or {2, 4} with multiplication. But {2, 6} is already used in the [12] cage in the fourth row, so the bottom cell in the [8] cage can't be 2 or 6. Hence, it must be 4, with the 2 above.

- The last two cells in the sixth column must be {1, 2}, which means the other cell in the [6] cage must be a 3; the operation can be either addition or multiplication. Further, the 2 in the [10] cage dictates the position of the {1, 2}.

- Similar logic applies to the [6] cage in the upper left, with {1, 2} completing the top row. Hence, the other cell in that cage must be a 3, and the order of the {1, 2} in the top row is dictated by the {2, 4} in the [8] cage.

There are still a lot of unfilled cells in the center of the puzzle, but our hard work has resulted in good progress. Twenty-one of the cells are now filled.

6 1	2	17 6	5	240 4	3
3	2 		6	11 5	4
8 2	11 			6	5
4	90 			12 2	6
15 			4 		6 2
		10 5	2	3	1

Much of the rest of the puzzle can be filled by identifying the missing values in a row or column.

- The [15] cage must have {5, 6} in the first column and {4, 6} in the bottom row, so the 6 must occupy the lower left cell. The placements of the 4 and 5 are then determined.

- The [2] cage in the second row must contain {1, 2}, with the placement dictated by the 2 in the [6] cage above.
- After the 1 is placed in the [2] cage, the three unfilled cells remaining in the second column must contain {3, 5, 6} in some order, and the 6's and 5 already placed in the third and fourth rows imply their placements.

With the exception of the [4] cage in the fifth row, every cage is now at least partially completed.

6 1	2	17 6	5	240 4	3
3	2 1	2	6	11 5	4
8 2	11 3			6	5
4	90 5			12 2	6
15 5	6		4 		6 2
6	4	10 5	2	3	1

Just seven cells left to go! A final bit of sleuthing will allow us to complete the puzzle.

- The [4] cage can be filled with either {1, 3} by addition or {1, 4} by multiplication. The important realization is that there must be a 1 somewhere in the cage.
- With a {5, 6} already in the [90] cage, the other two cells must be {1, 3}. Because a 1 must occur in the [4] cage, the bottom right cell of the [90] cage must be 3, with the 1 above it.

- Consequently, the last unfilled cell in the third column must be a 4, and the two remaining cells in the [11] cage must be {1, 3}.
- With {1, 3} filling the two cells in the fourth column of the [11] cage, the position of the {1, 4} in the [4] cage is now determined.

Finally, we're ready to place the last of the values and complete the puzzle.

6 1	2	17 6	5	240 4	3
3	2 1	2	6	11 5	4
8 2	11 3	4	1	6	5
4	90 5	1	3	12 2	6
15 5	6	3	4 4	1	6 2
6	4	10 5	2	3	1

And that does it! That wasn't so bad, was it?

Of course, the solution that appears above was written *after* the puzzle was solved, which makes it look like the solution was effortless. Far from it! To find that elegant solution required stops and starts. You'll experience the same thing when you solve some no-ops puzzles on your own, but be persistent! If you continue to explore the cages and consider possibilities, you'll eventually find something that works, and unlocking one door will likely unlock others.

Twist Puzzles

C'mon, baby, let's do the KenKen Twist

In a typical KenKen puzzle, the numbers 1 through n are used in an $n \times n$ grid. That is, the numbers 1 through 3 are used in a 3×3 puzzle, the numbers 1 through 4 are used in a 4×4 puzzle, and so on. But in KenKen Twist, you'll use a different set of numbers. Those numbers will be identified above the grid. For example, in the 4×4 KenKen puzzle below, instead of using 1, 2, 3, and 4, the numbers to be used are 2, 4, 6, and 8.

2,4,6,8

48×		8+	
12+		4−	
	8		20+
2÷			

Otherwise, the regular rules of KenKen still apply.

Hints for KenKen Twist

To solve KenKen Twist puzzles, you'll follow the same process as solving standard KenKen puzzles, but the following hints may prove helpful:

- If you don't know which number goes in a cell, note the possibilities! Use what you know to eliminate as many possibilities as you can. (This is the first and most important rule for solving any KenKen puzzle: standard, no-ops, twist, or otherwise.)

- Note the factors of the numbers in the candidate set. For instance, for the set {2, 4, 6, 8}, the number 6 is the only one with a factor of 3. (The other three numbers are all powers of 2.) This is important; you will know immediately if a multiplication cage contains a 6 or not, based on whether the target number is a multiple of 3. If the target number is a multiple of 9, then the cage would contain two 6's.

- Look for special numbers in the candidate set, such as prime numbers. If 5 and 7 appear in the candidate set, they'll provide a lot of information, especially regarding multiplication cages.

- If one number in the candidate set is significantly larger than the others, it's a strong possibility to be involved in large sums. For instance, if a [25+] cage has just four cells, and the candidate set is {1, 3, 4, 9}, it's a sure bet that at least one 9 will appear in the cage.

- Consider the collection of numbers in the candidate set. For example, the candidate set {2, 3, 5, 7} contains only prime numbers, so the multiplication and subtraction cages will be unique; addition cages will contain at least one 2 if there are an odd number of cells and the target number is even; and, division is impossible. Similarly, the candidate set {1, 2, 4, 8} contains only powers of 2, which means there will be many possibilities for multiplication and division, but subtraction will have unique target numbers; and the candidate set {2, 4, 6, 8}, which was used for the puzzle above, contains only even numbers.

A KenKen Twist example

To solve the KenKen Twist puzzle below, the first step is to fill in the [8] cage. That proves helpful, because the [48×] cage must be filled with {6, 8}, and since 8 cannot be repeated in the second column, it leads to this:

2,4,6,8

48× 8	6	8+	
12+		4−	
	8 8		20+
2÷			

The two unfilled cells in the top row must be {2, 4} in either order. Since their sum is 6, the remaining cell in the [8+] cage must be a 2. But that implies that the last cell in the first row cannot also be a 2, which yields:

2,4,6,8

48× 8	6	8+ 2	4
12+		4−	2
	8 8		20+
2÷			

The bottom two cells in the fourth column must then be filled with {8, 6}; and the [8] cage dictates their order. Further, the bottom cell in the third column must be 6 to complete the [20+] cage:

2,4,6,8

48× 8	6	8+ 2	4
12+		4−	2
	8 8		20+ 6
2÷		6	8

Similar logic allows the rest of the puzzle to be completed:

- The [4−] cage must be filled with {4, 8}.
- The [2÷] cage must be filled with {2, 4}.
- The [12+] cage must be filled with {2, 4, 6}.

The numbers already filled dictate the placement of those remaining to give the final solution:

2,4,6,8

48× 8	6	8+ 2	4
12+ 6	4	4− 8	2
2	8 8	4	20+ 6
2÷ 4	2	6	8

That wasn't so bad now, was it?

A larger KenKen Twist example

A 4×4 puzzle only has 16 cells. Let's see how the complexity changes for a 6×6 twist puzzle with 36 cells. Buckle in!

Instead of the numbers 1 through 6, the 6×6 KenKen Twist puzzle below uses the candidate set {1, 2, 3, 7, 8, 9}. Fill in the cages with the values 1, 2, 3, 7, 8, or 9 so that the given operation yields the target number. As always, you may only use each number once in each column and row.

1,2,3,7,8,9

18+		21+	4÷		7×
			4×		
5−				27+	
8−	14×	5−			
		15+		1728×	3
2÷					

Before you begin, consider the candidate set. You can think of it as bimodal, since three of the numbers (1, 2, 3) are small, and three of the numbers (7, 8, 9) are big. There is a gap between these two subsets. That may be important when solving the puzzle, since 4, 5, and 6 won't be used.

In addition, take note of those cages in the puzzle that must be filled with unique numbers:

- Of course, the [3] cage will be filled with 3.
- The [2÷] cage in the last row must be filled with {1, 2}, because none of the three large numbers (7, 8, 9) in the candidate set yield a quotient of 2 when divided by any of the three small numbers (1, 2, 3).
- The [4÷] cage can only be filled with {2, 8}.
- The three-cell [4×] cage can only be filled with {1, 2, 2}.
- The [7×] cage must be filled with {1, 1, 7}.
- The [14×] cage must be filled with {2, 7}.

- The [8–] cage must be filled with {1, 9}. (This is equivalent to a [5–] cage appearing in a standard 6×6 KenKen puzzle.)

- Finally, the [1728×] cage is equal to $12 \times 12 \times 12$, which is equal to $2^6 \times 3^3$. This is an important point: to get six factors of 2, two 8's will be needed; and then to get three factors of 3, a 3 and a 9 will be needed. Consequently, this cage will be filled with {3, 8, 8, 9}, which means that an 8 will appear in the top cell of the cage.

Based on these observations alone, we're able to fill 11 cells in the puzzle.

1,2,3,7,8,9

18+		21+	4÷ 8	2	7× 1
			4× 2	1	7
5–		2	1	27+	
8–	14×	5–			
		15+		1728× 8	3 3
2÷					8

That's an auspicious beginning!

There are now some additional observations that can be made:

- In the sixth column, the two remaining cells must be {2, 9}, since the other four numbers have been used. Then, the other two cells in the [27+] cage must be {7, 9} to get the required target number.

- Since {7, 9} will be used in the fifth column for the [27+] cage, then the remaining {3, 9} in the [1728×] cage cannot have the 9 in the fifth column.

- Since {1, 2} will be used in the [2÷] cage, and since {3, 8, 9} are used in the sixth row as part of the [1728×] cage, then the remaining cell in the bottom row must be 7. Consequently, {1, 7} must be used in the other two cells of the [15+] cage, and their order is dictated.

Adding in these additional pieces gets us more than halfway home, and we're well on the way to a full solution.

1,2,3,7,8,9

18+		21+	4÷ 8	2	7× 1
			4× 2	1	7
5–		2	1	27+ 7	9
8–	14×	5–		9	2
		15+ 1	7	1728× 8	3 3
2÷		7	9	3	8

Because of the pieces that are already filled in, the order of the values in the [8–], [14×], and [2÷] cages are dictated and can be filled in:

- The [8–] cage must be filled with {1, 9}, and the 9 cannot be in the top cell.

- The [14×] cage must be filled with {2, 7}, and the 7 cannot be in the bottom cell.

- The [5–] cage in the fourth row must be filled with {3, 8}, because a 2 has already been used in that row. Further, the 8 in the [4÷] cage of the first row dictates their order.

- The [2÷] cage must be filled with {1, 2}, and the {2, 7} in the [14×] cage dictates their order.

That's a lot of information, and filling in those values means the puzzle is three-fourths complete.

1,2,3,7,8,9

18+		21+	4÷ 8	2	7× 1
			4× 2	1	7
5–		2	1	27+ 7	9
8– 1	14× 7	5– 8	3	9	2
9	2	15+ 1	7	1728× 8	3 3
2÷ 2	1	7	9	3	8

Finally, the following can be noted:

- The numbers {3, 9} must fill the remaining cells in the third column, so the [21+] cage must be filled with {3, 9, 9}.

- That leaves {3, 7, 8} to fill the [18+] cage.

- The only candidates remaining for the [5–] cage in the third row are {3, 8}.

The placement of these values is determined by the other numbers in each row and column, so this solves the puzzle.

1,2,3,7,8,9

18+ 7	3	21+ 9	4÷ 8	2	7× 1
8	9	3	4× 2	1	7
5– 3	8	2	1	27+ 7	9
8– 1	14× 7	5– 8	3	9	2
9	2	15+ 1	7	1728× 8	3 3
2÷ 2	1	7	9	3	8

Do not be fooled! The step-by-step solution above makes it appear that the solution was easily found, but it wasn't! Before a concise, lucid solution can be obtained, many false starts and mistakes may occur. Do not be discouraged if you have trouble solving KenKen puzzles or if your solution process isn't as smooth as the solution above. That's normal! Remember, you're a rock star for even trying these puzzles, and the point is to have fun, improve your problem-solving ability, and develop patience. Don't stress. Enjoy!

4×4 PUZZLES

4×4 Lim-Ops Puzzles (Addition & Subtraction)

Let's start with 4×4 addition and subtraction puzzles. The goal is to fill each cell using only the numbers 1, 2, 3, and 4 without repeating a number in any row or column.

> ***Tip:*** *Remember, a "freebie" is always a great place to start. In Puzzle 1, fill in the "2" near the bottom right corner first, and you're on your way!*

1

6+	1−	6+	
			7+
2−		2	
3−		1−	

SOLVING TIME

2

4+		2	2−
9+	3−		
		2−	
1−		1−	

SOLVING TIME

Tip: *Remember, numbers can be arranged in any order within a cage when using subtraction.*

3

17+			
2–		2–	
	7+		3–
1			

SOLVING TIME

4

8+		7+	
	3		2–
7+		3	
	1–		3

SOLVING TIME

Tip: *Sometimes, you might know which numbers have to go in a cage, but not know the order (2+3 or 3+2?). Mark your options down in pencil and come back to them when you've filled in more of the numbers around them.*

5

3+		9+	5+
3			
3–		8+	
1–			

SOLVING TIME

6

10+	1–		10+
5+		4	1–
	3–		

SOLVING TIME

7

3+	11+		
	12+	3−	
			2
	4	1−	

SOLVING TIME

8

9+		1−	
3			2−
3−	4	6+	
			2

SOLVING TIME

9

8+		3−	
3−		2−	
	8+	8+	

SOLVING TIME

10

2−		9+	
2−	8+		
	2		3−
1	1−		

SOLVING TIME

Tip: *When you see an I-shaped cage of 3 squares that uses addition in a 4 × 4 puzzle, use deductive logic to figure out the 4th square in that row or column. The sum of the entire row or column (and ANY row or column in a 4 × 4) MUST be 10 (1 + 2 + 3 + 4 = 10). If the target number in a 3-square I-shaped cage is [9+], the one square left in that row or column must be a 1 (since 10 − 9 = 1).*

11

2−		2−	
9+	3−		1−
	1−		
	1	1−	

SOLVING TIME

12

5+	9+	1−	
			1−
8+	3−		
		3−	

SOLVING TIME

13

2−		7+	
3−	1−		
	2−		8+
5+			

SOLVING TIME

14

7+		10+	1−
2			
		10+	
2−			

SOLVING TIME

15

1−	1	6+	
	8+		
10+			3
		5+	

SOLVING TIME

16

2−		6+	
2−	7+		
	4		1−
6+			

SOLVING TIME

17

9+	3−		7+
	6+		
		9+	
2−			

SOLVING TIME

18

6+		2	1−
	8+		
2		6+	
1−			2

SOLVING TIME

19

1−	8+		
	9+		3+
8+		1−	
			4

SOLVING TIME

20

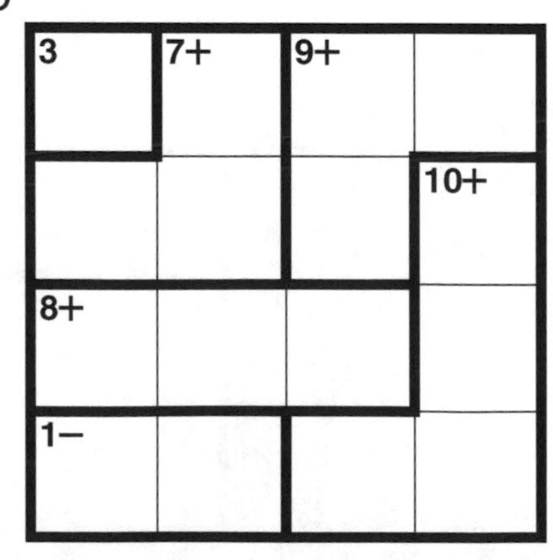

SOLVING TIME

4×4 Lim-Ops Puzzles (Multiplication)

21

24×	4	6×	
		3	
	12×	8×	
2			

SOLVING TIME

22

12×		24×	
6×			
8×			4
		6×	

SOLVING TIME

23

12×		12×	
4			
6×	2	48×	

SOLVING TIME

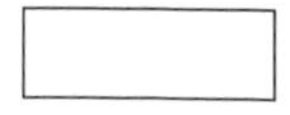

24

6×	4	24×	
			8×
12×			
	6×		

SOLVING TIME

Tip: *Take a few seconds to look over the full puzzle before you begin. It's a great way to quickly pick out some of the easier-to-solve cages.*

25

24×	6×		
		12×	
2	12×	8×	

SOLVING TIME

26

24×			24×
6×	2×		
	4		
	12×		

SOLVING TIME

27

3	24×		16×
8×		36×	

SOLVING TIME

4×4 No-Ops Puzzles (All Operations)

28

6		3	
2	9		2
	1		7
5			

SOLVING TIME

29

36	2		4
		2	2
8		9	

SOLVING TIME

30

2	3	36	2
8		8	
4			

SOLVING TIME

31

1		2	
7		2	12
2		1	
			1

SOLVING TIME

32

72	5		2
		2	
4		2	
3		12	

SOLVING TIME

33

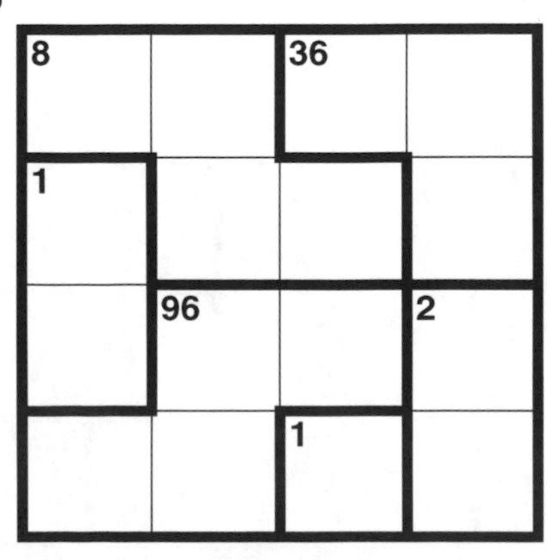

SOLVING TIME

34

2	3	4	1
2	13		
		1	

SOLVING TIME

35

36	2		3
2		1	1
	4		

SOLVING TIME

36

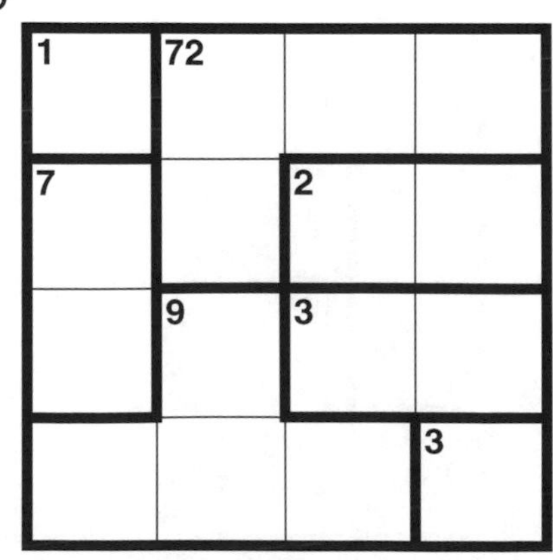

SOLVING TIME

37

12		1	
	10		1
2		2	8
	3		

SOLVING TIME

38

2	2		1
	5		
2	18		3
	1		

SOLVING TIME

39

2	3		6
	9	3	
5			
		2	

SOLVING TIME

SOLVING TIME

40

15			1
	2		
		10	
2			

SOLVING TIME

41

3	1	2	
		18	
2		5	
1			4

SOLVING TIME

42

4	6		
6	2		4
	1		24
	1		

43

2	12		6
	7		
9			
1		2	

SOLVING TIME

Tip: One thing we forgot to tell you ... there's no guessing necessary in KenKen, so don't bother guessing! Narrow down your options methodically until there's only one possible answer for each square. You can use this strategy for 4 × 4 puzzles, 5 × 5 puzzles, and even 9 × 9 puzzles (when you're brave enough to tackle them).

44

1	2		2
	1	32	
3			4
	1		

SOLVING TIME

SOLVING TIME

45

9		2	2
	10		
		12	
1			3

SOLVING TIME

46

3	2		9
24			
	2		3
		3	

SOLVING TIME

47

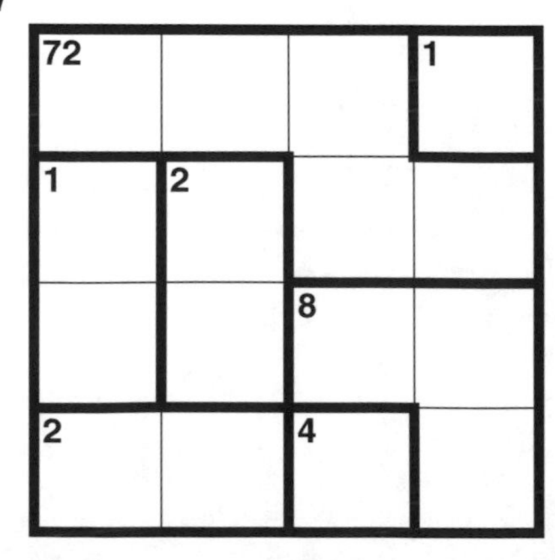

4×4 Twist Puzzles (All Operations)

48 *1,4,6,7*

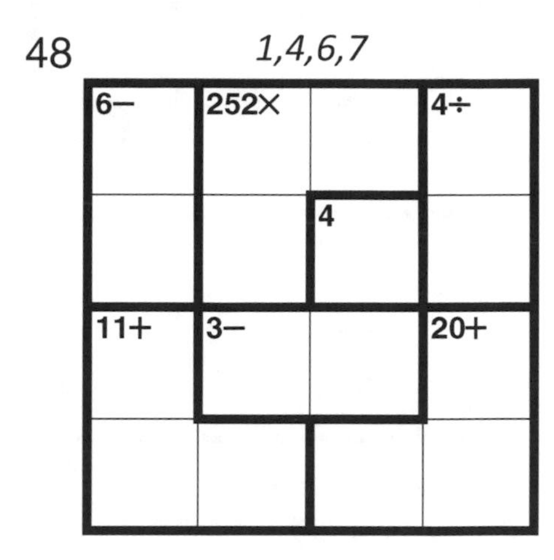

SOLVING TIME

49 *1,4,6,7*

6−		10+	5−
168×	4÷		
		28×	
	5−		7

SOLVING TIME

50 *1,4,6,7*

24×		12+	
5−	17+		
	1		42×
7	4÷		

SOLVING TIME

51 *2,3,5,8*

8+	5	4÷	
	6−		3
4÷	120×	3	3−

SOLVING TIME

52 *2,3,5,8*

2	13+		360×
4÷			
120×	3	3−	
			2

SOLVING TIME

53 *2,3,5,8*

15×	15+		
	15+	3	1−
4÷			
	2−		8

SOLVING TIME

54 *3,4,6,9*

12+		216×	
4	3−		3−
2÷		4	
	1−		9

SOLVING TIME

55 *3,4,6,9*

144×	3÷		108×
9+		22+	
3	2−		

SOLVING TIME

56 *3,4,6,9*

12×		15+	
2−	3−	3	3÷
		14+	
3÷			

SOLVING TIME

57 *4,5,6,8*

2÷	1920×	600×	
16+			
5		4−	

SOLVING TIME

58 *4,5,6,8*

2÷		30×	
17+	9+		2÷
		8	
40×		2−	

SOLVING TIME

59 *4,5,6,8*

11+	2−	80×	8
2÷		1440×	
1−			

SOLVING TIME

60 *1,3,6,9*

4374×			5−
	16+		
		1	18+
3÷			

SOLVING TIME

61 *1,3,6,9*

18×		3÷	
9			3−
21+		8+	

SOLVING TIME

62 *1,3,6,9*

18×	6	3÷	
			13+
27+			
		5−	

SOLVING TIME

63 *2,4,6,8*

256×

4−

10+

28+

3÷

2

4

SOLVING TIME

64 *2,4,6,8*

256×

4−

8+

28+

3÷

8

SOLVING TIME

65 *2,4,6,8*

2048×

3÷

16+

2

20+

4−

SOLVING TIME

66

1,2,6,8

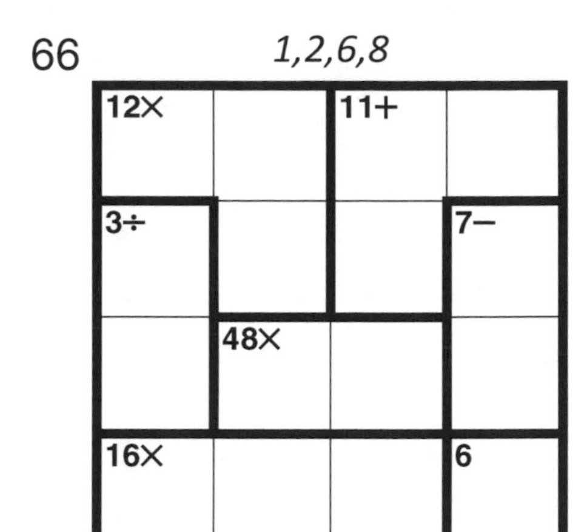

SOLVING TIME

67

1,2,6,8

9+	576×	2÷	6
			6−
2		48×	

SOLVING TIME

5 × 5 PUZZLES

5 × 5 Lim-Ops Puzzles (Addition & Subtraction)

68

3	7+	14+	9+	
			4–	
2–			7+	9+
7+	1–			

SOLVING TIME

69

11+			1	1–
1–	9+	4–		
		4	1–	9+
		13+		
3–				

SOLVING TIME

Tip: *As with baking, ping pong, and playing the bassoon, when it comes to KenKen, practice makes perfect. The more puzzles you do, the faster your brain will process the next one! Keep going ... you'll see exactly what we mean.*

70

3+ 14+
8+ 3– 2–
10+ 13+
6+
5+ 5

SOLVING TIME

71

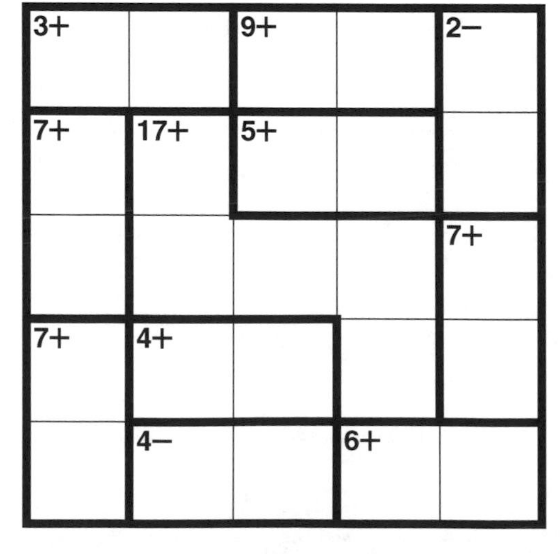

SOLVING TIME

72

7+	12+			
	4−		3	3+
6+	1−	2	11+	
				12+
6+				

SOLVING TIME

73

1−	4+	5	3−	2−
		1−		
3+	9+		3−	2−
		1−		
3−			4+	

SOLVING TIME

74

12+	4–		3–	4
	8+			1–
	12+			
3+				9+
	9+			

SOLVING TIME

75

7+	9+	13+		
		4–		
2	13+		9+	
4–				
	2	8+		

SOLVING TIME

***Tip:** As you try solving these puzzles, fill in possible answers in pencil (most people write small numbers at the bottom of the squares or on the lines) and erase them as you narrow down the possibilities.*

76

3	3+		4−	11+
4−	10+			
		14+		
6+				8+
8+				

SOLVING TIME

77

15+			1−	3−
3+		11+		
			3−	
9+			4−	1−
5+		2		

SOLVING TIME

78

2−		7+		1−
12+	5+		7+	
	5			11+
	15+			
2				

SOLVING TIME

79

9+		7+	4−	1−
15+				
	9+		2−	
			9+	3−
	2			

SOLVING TIME

80

11+		15+		
			12+	8+
1	1–			
3–	4–			
	7+		3+	

SOLVING TIME

81

12+	6+			5
		7+		14+
14+	10+			
	2			
			3–	

SOLVING TIME

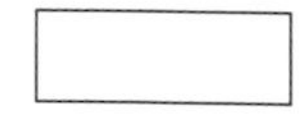

82

SOLVING TIME

16+				4
	7+		1	15+
2–		1	3–	
12+				
2	2–			

83

SOLVING TIME

5+		11+	6+	5
9+				1–
	12+			
1–				3–
	12+			

84

1−		8+		
1−	4−		13+	2−
	1−	12+		
3+				5
	1			

SOLVING TIME

85

7+		16+		
1−		12+		
3	11+			
4−		10+		6+
		1		

SOLVING TIME

86

9+		7+	9+	
3				14+
1−				
4−	8+	11+		3−

SOLVING TIME

87

8+		9+		5+
	2	8+	2−	
3−				3−
	8+		10+	
1−				

SOLVING TIME

5 × 5 Lim-Ops Puzzles (Multiplication)

88

15×, 12×, 40×, 12×, 8×, 60×, 5, 12×, 10×

SOLVING TIME

89

10×, 12×, 30×, 24×, 8×, 15×, 15×, 20×, 8×

SOLVING TIME

90

2	12×	40×		
		30×		24×
20×				
	30×		60×	

SOLVING TIME

91

30×		240×		
8×			3	
		5	80×	
20×	3	6×		

SOLVING TIME

92

20×		120×	15×	
				24×
10×		3		
120×				
	4		2×	

SOLVING TIME

93

120×				5
	10×	24×	12×	
			15×	
2×				120×
4				

SOLVING TIME

94

240×		3×	10×	
5			48×	
		40×		
	12×		5	
			6×	

SOLVING TIME

5 × 5 No-Ops Puzzles (All Operations)

95

18	9		1	2
		9		
7			8	
11		1		1
	2	3		

SOLVING TIME

96

2	2	20		7
9		75	16	
4	2			4
		1		

SOLVING TIME

97

1	9		1	
48		45	15	
	7			
12			2	

SOLVING TIME

98

2	9		2	2
	10	50		
720				1
			2	

SOLVING TIME

99

8		10		3
	7	9	3	2
90			2	
	4			13

SOLVING TIME

100

5	40		90	
12	1			
		48		
8			2	
		4		4

SOLVING TIME

101

2		6	4	
45			96	
		5		3
	14			
2			8	

SOLVING TIME

102

2	5	1		2
	9			
4		24		
	5		100	
7		2		

SOLVING TIME

103

2	4		36	
	11	6		5
3			4	
6			9	
	12			

SOLVING TIME

104

2	1	720		
	3	8	5	
20				
			6	
4			1	

SOLVING TIME

Tip: *Remember, you can use the same number more than once in an "L-shaped" cage, provided the identical numerals are in different rows or columns.*

105

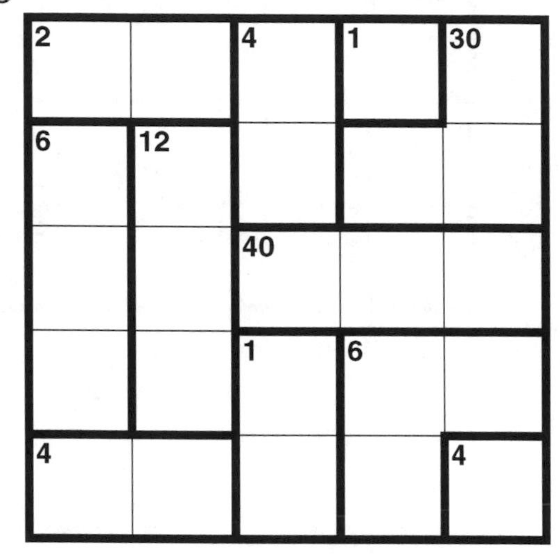

SOLVING TIME

106

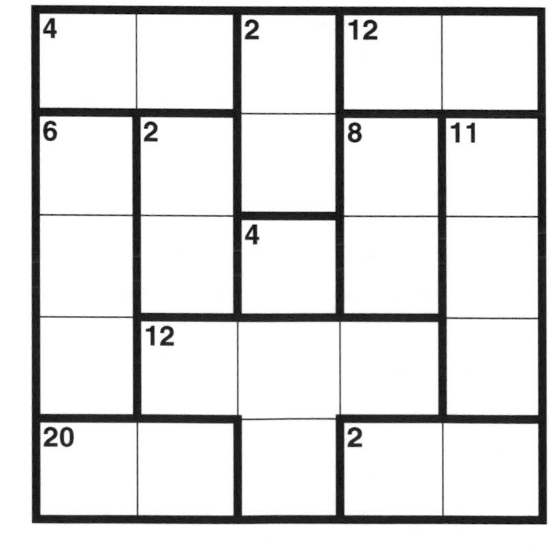

SOLVING TIME

107

15		2	13	
	12			6
		1		
7		4		
	24			5

SOLVING TIME

108

12	4		2	
	9		5	9
1		5		
	13		1	
		7		

SOLVING TIME

109

10		12		2
	2	15		4
1	5	4		
		2		10
5	1			

SOLVING TIME

110

2	5	2	75	2
9	1		1	11
	8			
	24			

SOLVING TIME

111

72			1	6
6		40		
	1		2	
	15		2	1

SOLVING TIME

112

20		2	2	
			4	2
7				
1	4		60	
	2			1

SOLVING TIME

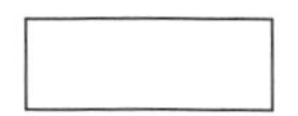

113

SOLVING TIME

1 1 50 1
3
16
15 12
2 1

114

SOLVING TIME

9 4 8
5 19
2 120
13

5×5 Twist Puzzles (All Operations)

115 *1,2,4,5,8*

3+		8×	160×	
7+			8	
	12+	3−	10×	
160×				2÷
		4÷		

SOLVING TIME

116 *1,2,4,5,8*

8	3+		20×	
4−	2÷	9+		64×
		20×		
8×			5	
	1−		7−	

SOLVING TIME

117

1,2,4,5,8

9+ | 7− | 10× | 4÷ | 640× | 4− | 12+ | 160× | 7+

SOLVING TIME

118

1,3,5,6,9

3÷ | 810× | 6+ | 4− | 270× | 5− | 10+ | 15+ | 81× | 4−

SOLVING TIME

119

1,3,5,6,9

90×	3÷		5–	
	4–	29+		3÷
			4–	
8–	2÷			1–
		2–		

SOLVING TIME

120

1,3,5,6,9

45×		18×		
7+		2–		21+
11+		10+		
3÷			1–	6+
	10+			

SOLVING TIME

121 *2,3,5,7,9*

2	19+	315×		
		15+	3−	
15+			54×	
	18×	3−	3÷	
				7

SOLVING TIME

122 *2,3,5,7,9*

441×		3−		3
	3−		3÷	
17+		378×	3−	
			28+	
1−				

SOLVING TIME

123

2,3,5,7,9

945×				2
4−	2	2−	16+	3÷
	8+			
11+		9		35×
	12+			

SOLVING TIME

124

4,5,6,7,8

11+		2÷		1−
336×	560×			
		27+	175×	
				192×
11+				

SOLVING TIME

125

4,5,6,7,8

210×			2÷	
1344×	100×	192×		7
				13+
		1−		
224×			1−	

SOLVING TIME

126

4,5,6,7,8

9+	192×			1−
	210×			
1−		2÷	20×	
192×	5		210×	
		1−		

SOLVING TIME

127

1,3,5,8,9

3÷	15×			7−
	22+			
4−		72×	3÷	
7−	3÷		18+	
				3

SOLVING TIME

128

1,3,5,8,9

675×		3÷	14+	
8−				11+
		216×		
5−				14+
	14+			

SOLVING TIME

129

1,3,5,8,9

45×	14+			24×
	13+			
15×		3÷	3240×	
5−				1
	3−			

SOLVING TIME

130

2,4,6,7,8

21+	2−		7	3÷
	3584×			
		11+		4
4			2688×	
2−		6		

SOLVING TIME

131

2,4,6,7,8

32×	1−		4−	
		25+	12+	
672×			4÷	
				294×
	6−			

SOLVING TIME

132

2,4,6,7,8

18+			5−	
2÷	1−		3072×	
	112×	3−		
		6		
112×			2−	

SOLVING TIME

133

2,3,6,7,9

378×		9	1−	
	2	2÷		4−
4536×	23+		36×	
				15+

SOLVING TIME

134

2,3,6,7,9

6	3÷		5−	
21×	2	3−		37+
	2÷	5−		
7−		9		
				6

SOLVING TIME

6×6 PUZZLES

6×6 Lim-Ops Puzzles (Addition & Subtraction)

135

9+		2	5−		10+
	4−	1−	11+	2−	
8+					
	6+	3−	12+		
			8+	3−	
11+				5+	

SOLVING TIME

136

6	12+	1–	11+	3–	
				2–	
	15+		1	15+	
		2			11+
3–		5–	5+	3–	
9+					2

SOLVING TIME

Tip: *Now that you've moved on to bigger puzzles with more possible numbers, don't get freaked out by cages with big target numbers! They might look intimidating, but often they only have one solution. Once you solve these, you can be confident you can conquer any KenKen puzzle ... all the way up to a 9×9.*

137

5+		2−	1−		11+
6+			5−		
5−	1−		10+		3−
	1−		4		
1−	10+		4−	2−	
	4−			5+	

SOLVING TIME

138

8+	11+	1−	2−		2
			9+		4+
	4−		6		
1−		2	4−	9+	
13+		5−		3−	
			12+		

SOLVING TIME

139

15+		6+		3−	
		5−		5+	
10+	5+		11+		9+
		1−	1−		
12+	3			7+	
		3−		4−	

SOLVING TIME

140

13+			1−		6
5−				3	9+
1−	3+		6	12+	
	13+		8+		
14+	4			5−	
		5+			

SOLVING TIME

141

3	23+		3+		13+
12+	5				
		5		3−	
15+			5	9+	2−
	5−	1−			
			8+		6

SOLVING TIME

142

19+					2
5+		1−		1	19+
5−	8+		14+		
		7+			
10+		5−		4−	
3		3−			

SOLVING TIME

143

5−	7+		19+		2
	7+		11+		
8+	5+			2	
	5−		2−	3−	5+
13+		3			
4		1−		4−	

SOLVING TIME

144

3+		2−	7+	17+	4
11+	2				
	12+		6	6+	
	19+			1−	
14+				7+	
			1−		

SOLVING TIME

145

2–	3–		14+		
	3–	3–	6+		
1			1–		5–
3–	12+		13+	4	
				1–	
9+			8+		4

SOLVING TIME

146

2–		10+		13+	5
5–	9+				
	20+	12+			5–
1–				7+	
					2–
5+		1	1–		

SOLVING TIME

147

2−		9+	15+		6+
6					
6+	1−		10+		
	1−	1−		5+	12+
		15+	7+		

SOLVING TIME

148

13+		4+		8+	
1−		25+	7+	2−	
	2−			5−	
1−					
	2	1−		8+	
9+		2	6+		

SOLVING TIME

149

22+					3
	10+	6+	15+	2	23+
5–					
		2	7+		
15+		6+			
3		5+			

SOLVING TIME

150

4	23+				
23+	9+		6+		
		4	3–		4–
	8+	4–	8+		
				4	11+
		3+			

SOLVING TIME

151

11+	4−		2−		4
		2−	8+	6+	
5−	11+			13+	2−
2−		5−	20+		
3−				4	

SOLVING TIME

152

18+		1−		1−	
	3−		9+		9+
	5−		1−		
		3	7+		
13+		8+		4−	9+
	3				

SOLVING TIME

153

12+			9+		
14+			11+		5+
1−	2		4−		
	9+			12+	11+
5					
14+			7+		

SOLVING TIME

154

20+			1−		1
	19+		13+		4−
1−					
		2	19+	15+	
8+	2				
		1−			

SOLVING TIME

6×6 Lim-Ops Puzzles (Multiplication)

155

360×
8×
180×
6×
40×
432×
5
24×
5
18×
4
60×

SOLVING TIME

156

40×
144×
18×
30×
80×
18×
6
10×
90×
20×
6×
48×

SOLVING TIME

157

90×			144×		
12×		8×			10×
		30×			
24×		6×	120×	15×	
12×					
	60×			2	

SOLVING TIME

158

12×		20×		6	12×
5		72×			
18×			12×		6480×
		16×	25×		
20×					
					2

SOLVING TIME

159

24×		20×		36×	
60×		6×			
		30×		48×	
	720×		3×		
6×				4	15×
			20×		

SOLVING TIME

160

20×	12×		36×		30×
		30×			
12×				720×	
	20×		2×		
18×		20×		4	
6×		6			

SOLVING TIME

161

1	4320×		30×		
17280×	1				
		180×		2×	
10×			12×		20×
		3			
		8×			3

SOLVING TIME

6×6 No-Ops Puzzles (All Operations)

162

11		2	15		
5	80			54	
					2
1	19			2	5
	8		21		

SOLVING TIME

163

100	12			36	
		5		3	
3	30		1		
	4			22	
108		7			
	3		5		

SOLVING TIME

164

11	1		12	3	24
	5				
	8	6		96	
3					50
		2	30		
24				3	

SOLVING TIME

165

6		800			90
	5		4		
1080	3	12			
			6	9	
				5	
11					1

SOLVING TIME

166

320			10		
	72		1		11
		10		3	
10	11		36		4
		5		12	
	3		4		

SOLVING TIME

167

3		180			
11		32		15	
18		11			2
	2		3	2	
16					1
		2		1	

SOLVING TIME

168

40	11		6		
			13		
12		11		2	
90		16	90	24	
		1		1	

SOLVING TIME

169

9		3		75	
5		6			
	900			13	
		32			1
1	11			7	
	2		5		

SOLVING TIME

170

3	40			3	
	2	5	10		100
1					
180		5		10	36
	25		16		

SOLVING TIME

171

5	1	80		13	
	3				
2	144			60	
60		3		3	
30		30			1
			1	24	

SOLVING TIME

172

8	5		60		
	40	16			
		15			
5		7		24	
108	2				2
		6			

SOLVING TIME

173

30			15		
3	2	4	8		
				1	
18	16		90		3
		4		2	6
			5		

SOLVING TIME

174

SOLVING TIME

2		1	30	1	
15				3	
6	40			1200	
		1			1
	10	3			
					4

175

SOLVING TIME

2		60	13		
1				13	
	7	2	5		
3			15		4
	17	14		2	

176

1		6	32	4	
2	5				13
	3		5		
1152	10		45		
			13		2

SOLVING TIME

177

2		3	800	18	
90	5				2
			10		
	3		3		1
8			2		
4		7		1	

178

288				14	6
	7				
		1	15	3	1
	144			5	
		5			4
5		2			

SOLVING TIME

179

144		6		6	
	2	1			22
2	24		120	90	
6		6			
14					

SOLVING TIME

180

2700	3				2
				4	7200
2	11	3			
40		13			3
144					

SOLVING TIME

181

5	2		80		
5	3			4	10
	960	72			
			4	90	
	4		72		
					1

SOLVING TIME

6×6 Twist Puzzles (All Operations)

182

1,2,4,5,6,9

14+		5−		8×	
5−		6	540×		
3÷	100×				4
	72×			2430×	
30×				2÷	
		2÷			

SOLVING TIME

183

1,2,4,5,6,9

432×					30×
11+		18×		30×	
3+		120×			15+
	8−		2÷		
11+				13+	2÷
	16+				

SOLVING TIME

184

1,2,4,5,6,9

180× 20× 3÷ 22+ 14+ 13+ 4− 6× 3− 1− 7+ 3− 8− 2÷ 5−

SOLVING TIME

185

1,3,5,6,7,8

15× 105× 14+ 6 20+ 1− 17+ 3 5− 1− 13+ 7− 6 3÷ 13+ 1−

SOLVING TIME

186

1,3,5,6,7,8

13+		4−		5−	
1	48×		2−		630×
840×	30×		15+		
	13+		20+		
	5−			6−	
		3÷			

SOLVING TIME

187

1,3,5,6,7,8

35×	3÷		14+		5
		24×	5−	1−	
30×	2−			4−	720×
		1−	12+		
19+				1	
		4−			

SOLVING TIME

188

2,3,5,6,8,9

1−		6480×	3÷		3
11+			10×	30+	
5		3÷		3÷	
15+	3−		5−		30×
	4÷		3÷		

SOLVING TIME

189

2,3,5,6,8,9

3÷	11+	48×	5−	2700×	
					4÷
6−	3÷	1−			
			28+		54×
30×	4÷	4−	3÷		

SOLVING TIME

190 *2,3,5,6,8,9*

9	3÷		360×		36×
2−	19+				
		8+		6480×	3−
108×			2		
	19+				54×
		45×		2	

SOLVING TIME

191 *3,5,6,7,8,9*

1−	3	75×	31+		
			21+		
35×		168×	9		
14+				14+	2−
2÷		3−	3−		
2−				11+	

SOLVING TIME

192

3,5,6,7,8,9

2÷		11+	3−	3−	27216×
2−					
11+	4−	13+			
			28+		
3−		45×	2÷	21×	
1−					

SOLVING TIME

193

3,5,6,7,8,9

7560×			22+		
3−	4−			14+	
		17+	24+		
2−	1−				40×
		972×	3÷		
			280×		

SOLVING TIME

194

1,3,5,6,7,8

3÷		40×	720×	2−	
21+					
	14+	21+			
		10+	2−	9+	
40×					2÷
	9+		42×		

SOLVING TIME

195

1,3,5,6,7,8

240×		17640×	3÷	7−	
				13+	
5−		200×	15+		6
				3÷	
21×		5−		5	16+
		42×			

SOLVING TIME

196

1,3,5,6,7,8

48×	5−		2−		4−
	13+		24+		
3÷		15+	6+		
2÷				21+	
525×					
	7	5−		11+	

SOLVING TIME

197

2,4,6,7,8,9

2	432×	1−		5−	
		17+	72×	165888×	3÷
1−					
	3−				9
13+					22+
	2÷		6		

SOLVING TIME

198

2,4,6,7,8,9

3÷		21+	2−		8
9	5−		96×	112×	
290304×					
	13+		22+		
	8		7−		3÷
				8	

SOLVING TIME

199

2,4,6,7,8,9

112896×				3÷	
4−	288×	1−		7−	
				17+	8
7		38+	3÷		36×
7−					
2−					

SOLVING TIME

200

1,4,5,6,8,9

10+	30×		2÷	5−	
	30×	1−		14+	
			2160×		8−
288×		7−			
14+			6+	9+	
	5−			5−	

SOLVING TIME

201

1,4,5,6,8,9

1	8−	120×		22+	
3−			288×	6	
	2−				13+
15552×	15+		4−		
	5			5−	
			4÷		5

SOLVING TIME

7×7 PUZZLES

7×7 Lim-Ops Puzzles (Addition & Subtraction)

202

9+			14+	8+		6
11+	17+				6–	
			3+		7+	2–
3+	10+			21+		
	13+				2–	
3–		5–		2–		1–
		2–		3+		

SOLVING TIME

203

12+		3+		17+	8+	6+
	18+	10+				
2–			6+		12+	
		12+				
4–	5+		4–		3–	
	4+		10+		7	1–
4	7+		11+			

SOLVING TIME

204

1–		14+			11+	
25+	4–		13+			6–
	4–			15+		
	15+				6+	6
	6–		17+			14+
	7+					
5–		11+				

SOLVING TIME

205

3+		17+			3−	
18+		5+		12+		
2−			5−		6	1−
	3−			5−		
18+			3−	6+		
10+	1−	5+		12+	10+	8+
			5			

SOLVING TIME

206

11+	8+		13+		7+	6−
		10+				
9+		19+		6−	15+	
						12+
6−	15+		6+	10+		
		14+		3		
11+					9+	

SOLVING TIME

207

2–	4–	19+				6
			1–		6–	
17+			9+	2	8+	
3+	2–				15+	
	6–	2–	13+	11+		
3–				2–		1–
	15+			3+		

SOLVING TIME

208

11+	2–	28+		1	1–	
		2–		7+		6
12+						5–
6+	2–		5	3–		
	27+			2–	6–	
	13+				3+	7+
11+		2				

SOLVING TIME

209

1−	5−		9+	3	7+	
	6−	17+		3−		3−
			6+			
			18+	28+		
1−					6−	
	2−		10+			2−
8+		4		7+		

SOLVING TIME

210

1−		4−		34+		2
9+					1	19+
	9+	6+		11+		
		11+	6−			2−
19+						
		3	12+			1−
1	3−			10+		

SOLVING TIME

211

13+			3	19+		
3+		6–		20+		2–
	10+		11+			
5	13+				2–	12+
13+		11+				
	15+					
1–		1	5–		11+	

SOLVING TIME

212

1–	24+		5+	17+		
	12+			3	2–	
6+			3–		7+	6+
			1–			
	8+			17+	2–	
3–						5+
	7	4–		9+		

SOLVING TIME

213

29+		4–		13+		6+
6		5–		7+		
3–		4–				
			3	5–	12+	1–
6–	13+		28+			
6+			15+			

SOLVING TIME

214

12+		5–		8+		2–
	7	16+	9+	21+		
3–						
	10+		7	8+		
13+			4–		16+	
5	24+					
4+						4

SOLVING TIME

215

16+		6+		20+	13+	
	15+	6+				
			3	8+		
6–	12+		3–		10+	
	6+			30+		
13+		9+			2	
		12+				

SOLVING TIME

216

1–		10+			19+	
9+	4–	3–	6–	13+		
						3–
	10+	10+	6	11+		
					35+	4
11+						5–
2			8+			

SOLVING TIME

217

18+		8+			5	6−
		19+		5−		
5			8+	24+		
6+					26+	
	4−	33+				3
					7+	
3−						

SOLVING TIME

218

11+	23+				14+	
			3+		23+	
2−		6−		16+		7+
2−			5			
4−		5+	1−			
17+				12+		12+
	9+					

SOLVING TIME

219

8+		18+	10+		14+	
				19+		
15+	8+					5+
	18+			7+		
	9+		5–		15+	
11+	3					16+
		11+				

SOLVING TIME

220

5–	7+	14+		6	1–	
				9+	1–	15+
4+	4–					
	15+	2–		4–		
2–			8+		12+	
	4	7+	11+			3–
11+					3	

SOLVING TIME

221

4	24+			14+		
		3−		3−		3−
6+	9+		11+	6+	1	
		4−			1−	4−
9+	4			6−		
	13+		22+			
11+						6

SOLVING TIME

7×7 Lim-Ops Puzzles (Multiplication)

222

21×		240×		72×		8×
	70×			1260×		
84×					15×	
			2			
1080×	1680×			12×		
				4×		245×
6×						

SOLVING TIME

223

840×			2×	12×		6×
	84×			15×	60×	
12×		120×				7
				42×		4×
30×			105×			
48×					168×	30×
7						

SOLVING TIME

224

60×		96×		210×		
		420×		126×		20×
7×						
		3	140×			126×
48×		360×			2	
126×				40×		

SOLVING TIME

225

28×		1050×	180×			
			48×	672×		
					6×	
18×	42×		5			
	504×			5	140×	
8×		15×		12×	7	
						6

SOLVING TIME

226

300×
140×
24×
42×
6
24×
630×
42×
20×
126×
96×
60×
28×
6×
105×
2

SOLVING TIME

227

30×
15120×
70×
18×
8×
7
48×
42×
4
120×
7350×
36×
120×
7
12×

SOLVING TIME

228

14×		7560×		16×	126×	5
		8×	30×		120×	
120×			7			252×
		126×				
60×		30×				56×

SOLVING TIME

7×7 No-Ops Puzzles (All Operations)

229

13	216	48			140	
			42	3		
				6	150	6
1	280					
		13		36		
24			19		7	3

SOLVING TIME

230

16		13		24	140	
11						6
		6	3	2	7	
					216	
13		75				336
	7			2		
14			6			

SOLVING TIME

231

180		168		8		3
	72				2	6
6	2					
		15			4	30
	2		7	24		
28		4	3	3		2
35				4		

SOLVING TIME

232

105	3		120			7
	17	140		3		
			1890		12	
1						11
2	4	2		1		
				1		147
3		20				

SOLVING TIME

233

2	84	2		90		
			6	96	3	
8		3				
6			14	4116		2
	360					
24			6			
		13			15	

SOLVING TIME

234

4		168				1050
3	3		9			
	270	13		70	7	
84			13			1
						15
	2		6			
	630					

SOLVING TIME

235

56 15 28 2
13 11 13
45 56
5 5 13 7
21
48 63 4
1

SOLVING TIME

236

2 20 3 15 2
11 35
108 6
168 2
13 1680 2 15
1 72
3 7

SOLVING TIME

237

336		20		2		5
		35		13		
3			24			
	5	27	11		13	
3	7			5		
			6		45	
3						4

SOLVING TIME

238

3	3		144		30	10
2	7	1				
	7	2			18	
20			70			
			6		216	2
4		70	4			
11						1

SOLVING TIME

239

5	1152	2		588		
		16		210		
	105				2	
			108			5
		15	6			
1176			2		1	8
			3			

SOLVING TIME

240

24		54	2		2	
	16			2	3	20
		1				
2		3		30		
	2		6	6		
2	12				6	
	240					7

SOLVING TIME

241

4	126			4	4	
	2	6		3		315
7		48	4			
	210			4		5
		2			2	
23	2		42			3

SOLVING TIME

242

50		2		3	5040	
	1008		2		10	
3						
	4		5	60		
2		30	18	7		
5880					4	
						6

SOLVING TIME

243

756		28	5250			
			8	2		
10		21				
	2				3	
30		6			2	
	2		168		42	6

SOLVING TIME

244

168		16				6
	2		2	12	6	
168		11			1	
			4		1	5
	24			7		
	30	6	13		72	
5			2			

SOLVING TIME

245

2 24 11 504
31
3 4 20 12
140 2 1
540 10
3 588
3

SOLVING TIME

246

45 29
6 18 40 6
1 2
4 35 4
14 100 2
3 756
11

SOLVING TIME

247

19 72 2 8
10
14 13 30 588 6
2 21
2
35 22 7 7
7

SOLVING TIME

248

6 1 24 2 1120
15
21 15 14
2 72
120 1 15 56
13 5 3

SOLVING TIME

7×7 Twist Puzzles (All Operations)

249 *1,2,3,5,6,7,8*

5	288×		6–		4–	
		6–	8+	14+		9+
	840×			7+		
8+				13+		
	1–	5–	126×		10+	3–
3÷						
	6–			15×		

SOLVING TIME

250 *1,2,3,5,6,7,8*

40×			2–	11+	27+	
6–	2÷	1–				
			6+			
3	84×			40×		
13+	2–	7–	4÷		70×	
			11+	84×		
90×						

SOLVING TIME

251

1,2,3,5,6,7,8

42×				14+		5
120×		56×		12×	2÷	
	16+		19+			3+
2−				14×		
	15+			2−		18+
36×			1680×			
	1−					

SOLVING TIME

252

1,3,4,5,6,8,9

24×		17+	6+		15+	
5−			15+			
180×		10368×				6×
3÷		180×				
	14+				2−	
17+	4÷		60×		576×	9
		3÷				

SOLVING TIME

253

1,3,4,5,6,8,9

8−	15×		14+		36×	
		26+		192×		
			4−	8+	17+	19+
4	7−					
30+			9			
1−	216×			54×		
			1−		3÷	

SOLVING TIME

254

1,3,4,5,6,8,9

288×			20×		144×	
2÷	15+					32+
		3÷	5	5−		
24×			17+		360×	
	4	14+	3−			
17+					1	
	2880×					1

SOLVING TIME

255

2,4,5,6,7,8,9

3–	3456×					2–
	70×		30×	20+		
2÷					2–	
210×			1134×	2÷		2
3–	1–				3÷	1–
		14+		2–		
3–		2–			13+	

SOLVING TIME

256

2,4,5,6,7,8,9

2÷	84×		405×		29+	2÷
		2–				
23+	3–					4–
	486×	224×			5	
			26+			315×
2–	17+	4÷		14+		

SOLVING TIME

257

2,4,5,6,7,8,9

3−		3−	17+	2268×		6−
1−						
17+	280×			5184×	2−	
					441×	
4÷			120×			
126×	216×		224×		1−	
				2	3−	

SOLVING TIME

258

3,4,5,6,7,8,9

3780×			1680×			
	4	18+	2−	17+		6−
				1−		
15+			2÷		27+	
15+	4−	3−	63×			
			60×		11+	
432×			7		7+	

SOLVING TIME

259

3,4,5,6,7,8,9

2÷		2−		288×		
14+	45×		5−		22+	
	4	18+	13+			270×
	17+		224×	5		
17+					180×	
	2−		2÷			
11+		3÷		19+		

SOLVING TIME

260

3,4,5,6,7,8,9

4032×		40×		16+		2−
	3	17+		1−		
		27+			3÷	
14+			16+	72×		7
	20+	6			35+	
45×			13+			
	4−		2−			

SOLVING TIME

261

1,3,4,5,6,8,9

240×		2−		13+		8
	12×		2−		54×	
3÷		120×		3−		
	2−		9	24×		1−
4−		8−	14+		108×	
	1152×					18×
			1−			

SOLVING TIME

262

1,3,4,5,6,8,9

3÷	4−		7−	12×	19+	
	48×					1−
3−		14+	162×	19+	5	
1−					7−	1
3−						2÷
9+	1−		20×	8−	216×	
	11+					

SOLVING TIME

263

1,3,4,5,6,8,9

17+	144×		3÷	26+		6+
					24×	
9+	36×					23+
	12×		3−		9	
	48×			12×	6+	
20+		225×	17+			1−
				4÷		

SOLVING TIME

264

1,2,4,5,6,7,8

140×			40×		9+	
5−		96×		14+	11+	
			120×			21+
9+				6−		
	24×		70×		2−	
240×	1−			2÷	24×	
						2

SOLVING TIME

265

1,2,4,5,6,7,8

42× 2− 3− 40× 7−
16+ 28×
1− 840× 6− 25+
6
160× 13+ 4÷
48× 1− 4− 2÷
2−

SOLVING TIME

266

2,3,4,5,6,7,8

112× 9+ 180× 2÷ 1−
30× 10+
6720× 336×
1− 14× 8
14+ 4− 11+ 960×
336× 3−
1−

SOLVING TIME

267

1,2,3,4,6,8,9

27×		2−		6−		54×
	14+		48×	12×		
48×		5−		1−		
	4÷				19+	5−
2÷		3÷		8−		
	9	6×	17+			20+
3						

SOLVING TIME

268

1,2,3,4,6,8,9

72×				2−	20+	
3÷	24×		16×			
	6−			48×		96×
1		5−			13+	
16+			10+			
	15+			21+	4÷	12×
4÷						

SOLVING TIME

8 × 8 PUZZLES

8 × 8 Lim-Ops Puzzles (Addition & Subtraction)

269

35+			11+	11+			6+
10+				5+	11+		
7−					13+		15+
5+	10+		12+			13+	
		6+		7−			1−
15+		15+	7−		1−		
				17+	15+		2−
	14+					2	

SOLVING TIME

270

15+		9+		27+			
4			9+		1−		
1−	14+			3+	1−	14+	
		7+					3
20+		7−		3−		2−	3−
6+		13+		18+			
		5−	7			7+	
5+			15+		15+		

SOLVING TIME

271

7	28+		8+		17+	14+	
5+	4						7−
		2−			5	21+	
8+			17+				3−
	16+	12+		7−		10+	
			7−	5+			
19+		8		13+		17+	
	1−		3+				

SOLVING TIME

272

27+		1		8+		33+	
3−							8+
	7−	3+		10+		1−	
11+		5+		2−			4−
	12+		7−		6+		
3−		1−		4−		6−	
9+		13+		7−		12+	
	2−		2−		3+		

SOLVING TIME

273

8+	13+	1−		3+		22+	12+
			1−				
14+	5−	7−		10+			32+
		7+				7	
	15+		3−				
	7−		21+		14+		
13+		3−			4−	5+	
10+			1−			4+	

SOLVING TIME

274

4+		3+	15+		1–		19+
30+			7+				
	6+			1–		15+	
			6	7–	4–	4–	
23+		18+				14+	
	8			1–		9+	
		1–	6+		7–		4+
3–			1–		1–		

SOLVING TIME

275

6+		12+		5–		23+	
31+		7+		7–			
	3	7–		1–	15+		
					15+	10+	5+
26+		18+	12+				
					14+		
		12+		4–		7–	5+
3–		13+		7+			

SOLVING TIME

276

13+	15+	12+	6+		7−		15+
				8+	16+		
6+		3+			6		
18+			3−		2−		11+
	1	3−	7−	41+	4−	14+	
	16+						
7+			7		10+	4+	
						3+	

SOLVING TIME

277

15+	11+		8	3+		1−	
	27+					1−	20+
	3+		14+	7+			
3+		4−		7−	2−		
18+						15+	
	5−	21+			4−		
		1	16+				1−
4−		12+			5+		

SOLVING TIME

278

17+		9+	3−	4−		1−	4+
7−				8+	9+		
	20+					15+	8
17+							15+
	5+	21+					
		1−	14+	1	7−	13+	6+
5+	13+			13+			
		4−			4		

SOLVING TIME

279

2−	4+	2−	21+			9+	
			5+		17+		2−
11+	7−			17+			
	6+			13+		15+	
7−	21+				9+		
	2−		2−			10+	7−
10+	11+	12+		6+			
						14+	

SOLVING TIME

280

13+		22+				7+	
6+		1–	7–		8+	21+	11+
	5		16+				
	6+			6–			6–
16+	7+			2–		9+	
	25+	9+	3	6+			5–
			11+		7		
		14+				3–	

SOLVING TIME

281

21+		8+		9+		7–	
			18+		21+	9+	
		6+	10+			4	15+
15+						12+	
	15+		21+	7–			10+
	18+					2–	
20+				17+	2–		9+

SOLVING TIME

282

12+	6+			19+		23+	
	9+	8+	5+		17+		
8+							2–
			14+	21+			
6–	7–				12+		9+
	17+		20+	9+			
21+					1–		
3		14+				3–	

SOLVING TIME

283

3–	7–		6+	7+	11+		13+
	8	3–			17+		
9+			4–			19+	
	12+		15+	21+	8+		
24+		6			3–		11+
		13+					
14+				7–			11+
		15+		7+			

SOLVING TIME

284

14+			21+			18+	
15+		6+					7–
	15+	12+		9+		7–	
			3–				18+
1–		8+	21+	5+	7–		
2–						8+	
9+	2–	21+			4	9+	4–
				8+			

SOLVING TIME

285

16+		1–		9+		1–	
	1–	10+		7+		10+	7–
21+		17+		16+			
	1–			8+		6+	16+
	7+	7+			12+		
		18+					
2–	19+		4–			8+	
		6–		16+			2

SOLVING TIME

286

17+		4−	14+	6+	11+		
	18+				19+	18+	
7−							
	8+		7+		13+	12+	
7+			1−			7−	
		13+		21+	6+	17+	
12+		5−					
3−		3−				3+	

SOLVING TIME

287

1−	2−	8+	7+	6−	5−	1−	
						15+	14+
7+	15+						
	9+	3−		29+		4−	
6−		14+				5+	
	18+					3+	1−
4−	15+	13+	4−	7+	5−		
						3−	

SOLVING TIME

288

3−	1−		14+		6−		2−
	6−	13+	16+		15+		
11+						5−	21+
	15+						
	1−		9+	5−	4−		
18+	2−				3−		11+
		4+		7+			
13+		14+			4	5+	

SOLVING TIME

8×8 Lim-Ops Puzzles (Multiplication)

289

840× 24× 24× 2352× 2×

90× 280×

192× 20×

16× 42×

168× 1200× 18×

28× 288×

12× 20× 24× 56×

5

SOLVING TIME

290

2880× 105× 18× 96×

210× 256× 336×

2

6× 20× 42×

320× 28×

12× 35× 60×

56× 24× 672× 12× 30×

SOLVING TIME

291

40× 4200× 2 96× 24× 28× 6 126× 6× 64× 30× 144× 280× 168× 6720× 6× 12× 10× 126× 224× 40×

SOLVING TIME

292

43200× 14× 1152× 7× 8064× 4 12× 600× 84× 576× 240× 2352× 7840× 5 16× 60× 6

SOLVING TIME

293

13440×				120×	24×		
	96×	84×					6
				168×	40×		14×
	42×		40×				
		5			6×	168×	
120×			32×			24×	
		288×	6×		35×		40×
			21×				

SOLVING TIME

294

3	768×		30×		24×	420×	
		336×					
			24×		56×	2880×	
30×		10×	84×				
	160×				392×		
		24×				6×	
	33600×				6		
42×							4

SOLVING TIME

295

5376×			280×		150×	12×	
	14×	12×					6×
		288×		112×			
				60×		336×	
2400×	20×				6×		
	24×		147×			192×	40×
				12×			
						42×	

SOLVING TIME

8×8 No-Ops Puzzles (All Operations)

296

7	2	15	24	30	4		42
	60				1680	21	
		10					
25	3					3	
			448		7	480	
84						48	
9		90	6	4			
					6		

SOLVING TIME

297

126	6		2		2		7
	768		1		1960		
			3			14	
3				15		1	
11		2016			3	420	
	3			2			
56		140			5	24	
5			48				

SOLVING TIME

298

90		42				2	
	2	192		15	7		11
15					4		
5			15	3		35	2
2		35			42		
6			144			288	
210					2		
		17				2	

SOLVING TIME

299

60	10	14		3	15		12
					1050	24	
	112						
15		1	5			2	
11				32		13	
7		280				4	120
	6		196	3			
6					2		

SOLVING TIME

300

2 2 15 26
30 12 2304 126 8
224 1
336 7 15
33 150 13 7
16
2

SOLVING TIME

301

63 100 5 64 2
3 1
9 144 13 5 60
15 2
20 3 12 1575
48 15 4 21
9
30 24

SOLVING TIME

302

2	3		21	72		12	40
	6				96		
42	28			7			3
	1				6		
7		12	24	15		36	
					1		
2	42	15		6	192		11

SOLVING TIME

303

896		3		9		9	
		7	84	180			
3	24					30	3
			10	15			
12		12			15		16
16				6			
17		7	56	2		6	
					84		

SOLVING TIME

304

270		20			12	3	15
		144		15			
3						200	
7	5	4	35	14			2
				45	11	2	
4	20	16					8
				6		90	
10		1					

SOLVING TIME

305

252				16		7	
8	3		2			126	
		14	105	14		3	
5				112	6		48
	2				3	5	
336		7				14	
		252	12		4		
2				6			

SOLVING TIME

306

144	3		3	14	4	19	
	1120	7					
		6			160		
		2		240	12	35	
	432		5			8	
	3					8	1
		14		30			
7					36		

SOLVING TIME

307

2	3	7		3		40	2
		3		4			
12		2		24		6	10752
2	60			4	3		
	21		21			8	
126			6				
	4		5		7		
		42		3			

SOLVING TIME

308

10080	28	120		7	7		6
			2		14		40
	4			224			
		3	240			13	
		2			1		10
60		6		12			
		29				28	
336							

SOLVING TIME

309

12	2		4	5	1		6
	8	4			40	3	
			5	20			2520
15	3		4		2		
	1						20
	15	7		2			
40		24	6				10
			2		2		

SOLVING TIME

310

3		21			336		5
			42		175		
54		8	2				5376
28		4		5	7	5	
	105			2			
4	30		10		4		
	7			9			2
5	7		28				

SOLVING TIME

311

75		1		11		1536	
	84	2	3	3	48		
8						1	2
		140	21	16800	6		
2						5	
7						1	
24			11			4	
	2						5

SOLVING TIME

312

2	2		336		1800	280	
				22			
36	1	3					
		4	1050				12
21	4		3		5		
	8		15				1
320				14	2		
5		3				3	

SOLVING TIME

313

9	448				1		5
	3		10		140		7
	210				48		
280	2			3		1680	
		4					
		2		13		7	
252	2	1	3				
			6		4	1	

SOLVING TIME

314

16	28			21	3	7	
						4	
1	3		42	11		7	
	56	8			11	7	
3			48			2	
	7			9		13	
96		28			3		3
	120				8	4	

SOLVING TIME

315

18		2		48	120	24	
15		3					4
	1		6		10	6	
		2	210				3
	5			4	1176		
		2	9				7
3			147	2	8		
2						9	

SOLVING TIME

9 × 9 PUZZLES

9 × 9 Lim-Ops Puzzles (Addition & Subtraction)

316

11+		16+			2−		2−	
2−	7+	1−		11+	8−		17+	4−
			16+		17+	5		
4−	14+					17+	5	23+
	19+						11+	
4−		8−	7+			3−		
			16+	17+				
14+	5−					1−		4+
	2−		9	4−		3+		

SOLVING TIME

317

18+		23+	4–		11+		3+	
	5		4+	15+		16+		
				8–	12+			21+
3–	3–		11+		2	12+		
	6–			16+		9	2–	
8+		24+	5		8+			14+
	8–		16+	4		5+		
11+				3–			3	16+
		13+			1–			

SOLVING TIME

318

8+	12+			2–	17+			17+
	1–		17+		8+	17+		
	3–			6+		3–		
4–		8+				2–	16+	
16+	8–		18+		4–		8+	9+
				14+				
13+		8–			3–		9+	
3–		20+			13+			8–
	7–		12+			13+		

SOLVING TIME

319

12+ 16+ 5– 14+ 8– 1– 12+ 3
4– 5–
23+ 6+ 11+ 18+
8–
28+ 11+ 8– 24+
3 19+ 3–
6+ 22+ 1–
1– 8– 8+ 4+ 2– 12+ 9+
6+ 12+

SOLVING TIME

320

30+ 3+ 12+
9+ 8+ 3+ 16+ 5– 8+
4– 15+ 17+
1– 10+ 9+ 2–
2– 25+ 8–
17+ 17+ 8– 16+ 1–
9+ 11+ 11+
5+ 3– 15+ 2
8+ 4– 3– 10+

SOLVING TIME

321

8– | 1– | 30+ | 3– | 7+ | 17+ | 7+ | 3– | 3– | 14+ | 5– | 6– | 2– | 5– | 20+ | 11+ | 4+ | 3– | 4– | 16+ | 3– | 8– | 19+ | 6+ | 13+ | 4 | 3+ | 20+ | 21+ | 1– | 22+ | 6– | 8– | 2– | 3+ | 7+

SOLVING TIME

322

7 | 22+ | 9+ | 19+ | 17+ | 7 | 9+ | 17+ | 9+ | 7 | 4– | 3– | 3– | 20+ | 8– | 10+ | 15+ | 6– | 6+ | 15+ | 8+ | 11+ | 17+ | 26+ | 9 | 22+ | 28+ | 8– | 15+ | 21+ | 1

SOLVING TIME

323

18+			14+		12+			18+
6+	4+	1−			8−		17+	
		7+		19+		33+		
3−			14+	7			4−	
16+	14+			2−			9+	
		8−		2−			11+	
1−	4−			17+			10+	6+
		9+		19+	13+			
18+						3+		

SOLVING TIME

324

9+		8−	11+		11+		16+	
3+	24+		4−		5	16+		
		5−		4	20+			3−
15+		5+		16+				
	16+			8−		8	17+	7+
13+		18+			5+	5−		
	8+			1−			10+	
14+	6		8−		6+			1−
		4		2−		7+		

SOLVING TIME

325

3– 10+ 16+ 5+ 8– 7+
21+ 1– 3 8+ 3–
4 6+ 4– 3–
4– 6 3– 14+ 4–
12+ 1 5+
8+ 5– 6+ 1– 1– 12+ 10+
13+ 14+
8– 1– 2– 9+ 11+
11+ 24+

SOLVING TIME

326

17+ 12+ 6+ 15+ 1– 1–
6– 2 1– 24+
2– 6+
2– 18+ 15+ 22+
1– 3– 7– 1
4– 30+
18+ 8– 12+ 23+ 11+ 3–
4+ 1–
23+ 4–

SOLVING TIME

327

12+	10+		1–		4–	16+	1–	
	10+	13+	1–					
9			16+	12+		20+		
10+					19+		15+	20+
		3	10+			15+		
	22+	8+		21+				
		8–				7+		
2–			1–		1–	15+		10+
	2–		17+					

SOLVING TIME

328

15+			2–		41+			
8–		2	2–	12+		11+		
5–		7+		8–		5		
14+			16+	9+		25+	4–	
	15+	19+		7				2–
			2	11+				
3–	17+		8–		10+		7	17+
			21+		11+	2–	5+	
7	1–							

SOLVING TIME

329

6–		1–		6+	12+	8–		21+
8–		19+				15+		
3+	18+		6+		18+	5		
				34+			11+	
17+	8–					10+		
			8–		2–		13+	5+
1–		13+		3–	12+			
21+	5		17+				5–	8–
					8+			

SOLVING TIME

330

13+	7	13+		10+				9+
	13+	18+	5–	7+	2–	30+	11+	
								18+
7–	8+		11+		3–		3+	
				24+				
4–		3–				2–	4–	
1–			20+				4–	
14+		8–		2–	13+	1–	4–	
3–			9				9+	

SOLVING TIME

331

17+			4−	4+	1−	12+		
20+	10+					4−		13+
	11+		11+			5−		
	15+		8−		1−	8+		
8+			5	3−			2−	
9+		12+			4−		2−	
9+		21+		15+	26+	11+	4−	
6−	4+						8−	4−

SOLVING TIME

332

18+		10+		3−		9+		
	11+	2−		30+				6−
12+			15+	8−	12+	13+		
	18+						2−	
		8−	14+	4−		5−	1−	
8+	19+				4+		2−	
				4		5+		13+
5−	21+			3−	1−		6−	
		1−				8		

SOLVING TIME

333

3–		15+			6+	11+	1–	5
7+	4–	2–	3+	6–				14+
					3–		17+	
11+	4	9+		23+				8+
		10+		2–			8	
8+		22+			17+	8–		
2–	11+					17+	2–	
		10+						2–
16+			11+			9+		

SOLVING TIME

334

16+	6+		8+		1–	11+		20+
	1–					5–		
	5–		16+			12+		
8+	6–		8–		15+	13+		5+
	16+	20+				21+	12+	
16+								8+
		7+		19+				
15+			19+			20+		
4+		12+			1–		7+	

SOLVING TIME

335

7−		17+	9+		24+		15+	
13+							10+	
13+	21+							11+
	3+		2−	16+	8+	14+		
14+	11+					13+		16+
			8+	1−	15+			
23+		6				6+	10+	
	13+	16+		8			23+	
			15+					

SOLVING TIME

9 × 9 Lim-Ops Puzzles (Multiplication)

336

2	7560×		288×		21×		40×	
		54×	5760×			56×		45360×
288×	14×		6	84×		120×		
					90×		3	
63×						64×		
	36×		105×					12×
20×		8			8640×		324×	
								7

SOLVING TIME

337

2160×			42×			5184×		
30×	735×		12×		72×		432×	16×
				56×				
					2400×			
864×							441×	
	96×			45×	84×			
	12×		864×			2250×		
		252×				112×		
			144×					

SOLVING TIME

338

60×		144×		6804×		168×		
135×				20×				56×
					32×			
1470×			12×		25920×		288×	
	48×			168×				
		1260×						
32×	432×			63×		150×	6×	
			27×		20×			
224×							18×	

SOLVING TIME

339

960×			1470×			9	162×	
	96×	2016×	2×					
				960×	84×		10×	5
4860×								
126×			24×				8×	
			9×		768×		126×	
56×	7	10×	29160×					
				60×		56×		288×
3								

SOLVING TIME

340

135× 252× 168× 8400× 108× 16× 4536× 80× 12× 48× 4032× 72× 45× 378× 160× 18× 8 75× 42× 672× 216× 189× 16× 40×

SOLVING TIME

341

40× 126× 48× 216× 180× 12× 504× 40× 504× 9 15× 14× 40× 56× 56× 144× 108× 4 225× 105× 24× 54× 12× 72× 504× 42× 20× 576×

SOLVING TIME

342

432×			2268×	84×			40×	
	480×					64×		
5		3024×					252×	56×
			6480×					
18×					75×		24×	
14×	4320×						336×	
		12×		84×	8			43740×
42×								
		80×						

SOLVING TIME

9 × 9 No-Ops Puzzles (All Operations)

343

4	3	280		8		17	16	
			224				60	
243		24		2880				
	32			21				22
280			108	5				
	336				10	3		2
	54		8	64			2	
9						3136	45	
		135						

SOLVING TIME

344

42		4		17		9072	20	
3		4	19					
160			6		12			
	8	5					126	12
14		2	280		180			
			1		3			15
5400			11			64		
224			216	21		6		4
6								

SOLVING TIME

345

15		56	120		8	10	2	17
17		168		7			150	
4			9072			17		
		162		300		18	56	
21							3	
30			4	1		3		252
22		18		3	2	5		
								4

SOLVING TIME

346

252				3		12	25	2
3	24	1920			63			
				4		432		
1960	12						4608	
	17		126			3		
		162		2	30			
13	180		8		4	294		
		7		8				
			70			17		

SOLVING TIME

347

1 384 3 567
336 252 9 108
42 3 4
17 4 80
6 2 270 168
15 12 1400 12
5 36 8
135 42 11 15
8 2

SOLVING TIME

348

192 21 6 2 25
17 80 90
8 3 1
15 19 26
80 48 3 189
45 12 30
105 4 23 6
2 63 64
42 13 8

SOLVING TIME

349

2 30 256 17
2 7 15 140 8
2 18 17
4 3 112
189 3 36 300
210 7 2 324
17 21
140 6 11
3 17 3

SOLVING TIME

350

2 3 22 24 14 9 14
441
2 17 4 60 5
17 1008 2
1 16 10
105 5 240 8 288
3 960 15
5 7 4 15 15
3 6

SOLVING TIME

351

105		1		14		18	2	
	9	70	7				20	
2					216			1
	32		5	567		36		
48					25	8		42
252		192						
			48	2		5	8	
8	8				2	105		25
		35		3				

SOLVING TIME

352

150		7			24	504		
	3	8	48	22			126	20
28	17							
			12			40		
	15		4	216			10	
4					8	378	4	
19	15		90				160	9
				2	6			6
	336					4		

SOLVING TIME

353

140		8	7	4		1	6	16
	6			96				
45	3				1	147		
	14	4	8				13	12
		320		20	5			
3					3	360		
	15		105	108				3
18		14				4	4	
				3				5

SOLVING TIME

354

2		10	63		19	5	5	
120			4				10	11
2		1	11	16	3			
	7					54	63	2
8	2	40		3	135			
			12			7	1	15
3	3	3					14	
		162		20		14		
7		4		42			2	

SOLVING TIME

355

4	14	17	96			8		
			210	10	3		20	
20					12			
	18	12	17		35	19		216
6								
12				360		3		
		63			3		2	
7	2	1			1			2880
			14					

SOLVING TIME

356

1280	4		8	3	45		2	
		15			14	72		4
567				3360				
		576			9	2	1	
1		8					2	
17						7	1	
	2	72			960			2
			3	8		4		
2	2				28		15	

SOLVING TIME

357

2		1	2	48		63	2	
13				36			3	
6		16				1		26
1			15		960			
1	19			2			84	
	200		17	10				
		35			6	3	5	
2	11		23	7				8
				16				

SOLVING TIME

358

30			17			90	7	27
720		30						
48		28	6	9			20	
					2			
7	3		7		7		3	
10			2	270	2		15	
	280					4		12
		42				1	14	
10				1				

SOLVING TIME

359

14		54	3		1		24	
2			30		240	5		
	10						17	36
9			2	17		147		
162	160	10		9				
		1	5		1	20		
			3	192				1
21	5			5		32		
		2		3			7	

SOLVING TIME

360

90		10	4		19	6	180	
			14	8			17	
15						9		5
	81		5	48			10	
3			4	360				6
7	3			105				4
		24		18	2		2	
576	100				2	6		56
			3					

SOLVING TIME

361

4	42	3		3	4		10	12
		5	162		13			
63						10		2
2		1		14				
10		24	21		9		8	13
	128			126		45		
			1		2	7	2	
6	13					1008		
		22					54	

SOLVING TIME

362

2160			63	20		14		
	240					7		
		4		7		84	48	27
126			160					
4		240		22		6		
			4		11		672	2
144	11			3		1800		
		7						12
		336						

SOLVING TIME

BONUS PUZZLES

No-Ops / Twist Puzzles (All Operations)

363 *1,2,3,6*

18			6
	3		8
9		1	
5			

SOLVING TIME

364 *2,4,6,7,8*

2688			4	
	19			17
	2	24		
1			6	
	56			8

SOLVING TIME

365

1,3,4,6,8,9

1152			20		
5				6	
5		27	18	1	
2	8			32	1
	14				
		2			9

SOLVING TIME

Ken-gratulations! You've conquered them all.

You are now officially a KenKen Master!

SOLUTIONS

1

6+ 2	1− 3	6+ 4	1
4	2	1	7+ 3
2− 3	1	2 2	4
3− 1	4	1− 3	2

2

4+ 1	3	2 2	2− 4
9+ 3	3− 4	1	2
4	2	2− 3	1
1− 2	1	1− 4	3

3

17+ 3	1	4	2
2− 2	4	2− 1	3
4	7+ 2	3	3− 1
1 1	3	2	4

4

8+ 3	4	7+ 2	1
1	3 3	4	2− 2
7+ 2	1	3 3	4
4	1− 2	1	3 3

5

3+ 2	1	9+ 3	5+ 4
3 3	2	4	1
3− 1	4	8+ 2	3
1− 4	3	1	2

6

10+ 4	1− 1	2	10+ 3
1	2	3	4
5+ 2	3	4 4	1− 1
3	3− 4	1	2

7

3+ 2	11+ 1	3	4
1	12+ 2	3− 4	3
4	3	1	2 2
3	4 4	1− 2	1

8

9+ 2	1	1− 3	4
3 3	2	4	2− 1
3− 1	4 4	6+ 2	3
4	3	1	2 2

9

8+ 3	2	3− 4	1
3− 1	3	2− 2	4
4	8+ 1	8+ 3	2
2	4	1	3

10

2− 3	1	9+ 4	2
2− 2	8+ 4	1	3
4	2 2	3	3− 1
1 1	1− 3	2	4

11

2− 1	3	2− 2	4
9+ 3	3− 4	1	1− 2
4	1− 2	3	1
2	1 1	1− 4	3

12

5+ 1	9+ 4	1− 2	3
4	2	3	1− 1
8+ 3	3− 1	4	2
2	3	3− 1	4

13

2− 3	1	7+ 4	2
3− 4	1− 2	3	1
1	2− 4	2	8+ 3
5+ 2	3	1	4

14

7+ 3	4	10+ 1	1− 2
2 2	1	4	3
1	3	10+ 2	4
2− 4	2	3	1

15

1− 4	1 1	6+ 3	2
3	8+ 2	4	1
10+ 1	4	2	3 3
2	3	5+ 1	4

16

2– 4	2	6+ 3	1
2– 3	7+ 1	4	2
1	4 4	2	1– 3
6+ 2	3	1	4

17

9+ 2	3– 1	4	7+ 3
4	6+ 2	3	1
3	4	9+ 1	2
2– 1	3	2	4

18

6+ 4	1	2 2	1– 3
1	8+ 2	3	4
2 2	3	6+ 4	1
1– 3	4	1	2 2

19

1– 2	8+ 4	1	3
1	9+ 3	4	3+ 2
8+ 4	2	1– 3	1
3	1	2	4 4

20

3 3	7+ 1	9+ 2	4
2	4	3	10+ 1
8+ 4	3	1	2
1– 1	2	4	3

21

24× 1	4 4	6× 2	3
4	2	3 3	1
3	12× 1	8× 4	2
2 2	3	1	4

22

12× 3	4	24× 2	1
6× 1	2	4	3
8× 2	3	1	4 4
4	1	6× 3	2

23

12× 1	4	12× 2	3
4 4	3	1	2
6× 3	2 2	48× 4	1
2	1	3	4

24

6× 1	4 4	24× 3	2
2	3	4	8× 1
12× 3	1	2	4
4	6× 2	1	3

25

24× 4	6× 1	3	2
3	2	12× 4	1
2 2	12× 4	8× 1	3
1	3	2	4

26

24× 4	3	2	24× 1
6× 3	2× 2	1	4
1	4 4	3	2
2	12× 1	4	3

27

3 3	24× 4	1	16× 2
1	3	2	4
8× 4	2	36× 3	1
2	1	4	3

28

6 3	2	3 4	1
2 4	9 3	1	2 2
2	1 1	3	7 4
5 1	4	2	3

29

36 3	2 2	1	4 4
4	3	2 2	2 1
8 2	1	9 4	3
1	4	3	2

30

2 1	3 4	36 3	2 2
2	1	4	3
8 3	2	8 1	4
4 4	3	2	1

31

(1) 3	4	(2) 1	2
(7) 1	3	(2) 2	(12) 4
(2) 2	1	(1) 4	3
4	2	3	(1) 1

32

(72) 3	(5) 1	4	(2) 2
2	4	(2) 1	3
(4) 4	3	(2) 2	1
(3) 1	2	(12) 3	4

33

(8) 1	2	(36) 3	4
(1) 2	1	4	3
3	(96) 4	2	(2) 1
4	3	(1) 1	2

34

(2) 2	(3) 1	(4) 4	(1) 3
4	3	1	2
(2) 1	(13) 2	3	4
3	4	(1) 2	1

35

(36) 3	(2) 2	4	(3) 1
4	3	1	2
(2) 2	1	(1) 3	(1) 4
1	(4) 4	2	3

36

(1) 1	(72) 4	3	2
(7) 4	3	(2) 2	1
3	(9) 2	(3) 1	4
2	1	4	(3) 3

37

(12) 4	1	(1) 3	2
3	(10) 2	4	(1) 1
(2) 1	4	(2) 2	(8) 3
2	(3) 3	1	4

38

(2) 1	(2) 2	4	(1) 3
3	(5) 4	1	2
(2) 4	(18) 3	2	(3) 1
2	(1) 1	3	4

39

(2) 2	(3) 4	1	(6) 3
4	(9) 2	(3) 3	1
(5) 1	3	4	2
3	1	(2) 2	4

40

(15) 1	3	4	(1) 2
4	(2) 2	1	3
3	1	(10) 2	4
(2) 2	4	3	1

41

(3) 4	(1) 3	(2) 2	1
1	4	(18) 3	2
(2) 2	1	(5) 4	3
(1) 3	2	1	(4) 4

42

(4) 4	(6) 3	2	1
(6) 3	(2) 2	1	(4) 4
1	(1) 4	3	(24) 2
2	(1) 1	4	3

43

(2) 2	(12) 4	3	(6) 1
4	(7) 2	1	3
(9) 3	1	4	2
(1) 1	3	(2) 2	4

44

(1) 2	(2) 3	1	(2) 4
3	(1) 1	(32) 4	2
(3) 1	4	2	(4) 3
4	(1) 2	3	1

45

(9) 1	3	(2) 2	(2) 4
3	(10) 4	1	2
4	2	(12) 3	1
(1) 2	1	4	(3) 3

46

3 3	2 2	1	9 4
24 4	1	3	2
1	2 4	2	3 3
2	3	3 4	1

47

72 2	4	3	1 1
1 4	2 2	1	3
3	1	8 2	4
2 1	3	4 4	2

48

6− 1	252× 7	6	4÷ 4
7	6	4 4	1
11+ 6	3− 4	1	20+ 7
4	1	7	6

49

6− 1	7	10+ 4	5− 6
168× 7	4÷ 4	6	1
6	1	28× 7	4
4	5− 6	1	7 7

50

24× 4	6	12+ 7	1
5− 1	17+ 7	6	4
6	1 1	4	42× 7
7 7	4÷ 4	1	6

51

8+ 3	5 5	4÷ 2	8
5	6− 2	8	3 3
4÷ 2	120× 8	3 3	3− 5
8	3	5	2

52

2 2	13+ 5	8	360× 3
4÷ 8	2	3	5
120× 5	3 3	3− 2	8
3	8	5	2 2

53

15× 3	15+ 2	8	5
5	15+ 8	3 3	1− 2
4÷ 8	5	2	3
2	2− 3	5	8 8

54

12+ 9	3	216× 6	4
4 4	3− 6	9	3− 3
2÷ 3	9	4 4	6
6	1− 4	3	9 9

55

144× 4	3÷ 9	3	108× 6
9	4	6	3
9+ 6	3	22+ 9	4
3 3	2− 6	4	9

56

12× 3	4	15+ 9	6
2− 4	3− 6	3 3	3÷ 9
6	9	14+ 4	3
3÷ 9	3	6	4

57

2÷ 4	1920× 8	600× 5	6
8	5	6	4
16+ 6	4	8	5
5 5	6	4− 4	8

58

2÷ 4	8	30× 6	5
17+ 6	9+ 4	5	2÷ 8
5	6	8 8	4
40× 8	5	2− 4	6

59

11+ 5	2− 6	80× 4	8 8
6	8	5	4
2÷ 8	4	1440× 6	5
1− 4	5	8	6

60

4374× 6	9	3	5− 1
3	16+ 1	9	6
9	6	1 1	18+ 3
3÷ 1	3	6	9

61

18× 1	6	3÷ 9	3
9 9	1	3	3− 6
21+ 6	3	8+ 1	9
3	9	6	1

62

18× 1	6 6	3÷ 9	3
3	1	6	13+ 9
27+ 6	9	3	1
9	3	5− 1	6

63

256× 4	8	4− 2	6
8	10+ 4	6	28+ 2
3÷ 6	2 2	4	8
2	6	8	4 4

64

256× 4	8	4− 6	2
8	8+ 4	2	28+ 6
3÷ 6	2	8	4
2	6	4	8 8

65

2048× 2	8	4	3÷ 6
8	16+ 4	6	2
4	6	2 2	20+ 8
4− 6	2	8	4

66

12× 1	6	11+ 8	2
3÷ 6	2	1	7− 8
2	48× 8	6	1
16× 8	1	2	6 6

67

9+ 1	576× 8	2÷ 2	6 6
8	6	1	6− 2
2 2	1	48× 6	8
6	2	8	1

68

3 3	7+ 1	14+ 2	9+ 4	5
4	2	3	4− 5	1
2− 1	3	5	7+ 2	9+ 4
7+ 2	1− 5	4	1	3
5	4	1	3	2

69

11+ 4	5	2	1 1	1− 3
1− 3	9+ 4	4− 1	5	2
2	1	4 4	1− 3	9+ 5
1	3	13+ 5	2	4
3− 5	2	3	4	1

70

3+ 1	2	14+ 4	5	3
8+ 5	3	2	3− 1	2− 4
10+ 3	1	13+ 5	4	2
4	5	3	6+ 2	1
2	5+ 4	1	3	5 5

71

3+ 1	2	9+ 5	4	2– 3
7+ 5	17+ 4	5+ 2	3	1
2	3	4	1	7+ 5
7+ 4	4+ 1	3	5	2
3	4– 5	1	6+ 2	4

72

7+ 3	12+ 2	4	1	5
4	4– 5	1	3 3	3+ 2
6+ 5	1– 3	2 2	11+ 4	1
1	4	5	2	12+ 3
6+ 2	1	3	5	4

73

1– 3	4+ 1	5 5	3– 2	2– 4
4	3	1– 1	5	2
3+ 1	9+ 5	2	3– 4	2– 3
2	4	1– 3	1	5
3– 5	2	4	4+ 3	1

74

12+ 3	4– 5	1	3– 2	4 4
4	8+ 1	3	5	1– 2
5	12+ 2	4	1	3
3+ 1	3	2	4	9+ 5
2	9+ 4	5	3	1

75

7+ 4	9+ 5	13+ 2	1	3
3	4	4– 1	5	2
2 2	13+ 1	4	9+ 3	5
4– 1	3	5	2	4
5	2 2	8+ 3	4	1

76

3 3	3+ 1	2	4– 5	11+ 4
4– 5	10+ 4	3	1	2
1	3	14+ 4	2	5
6+ 4	2	5	3	8+ 1
8+ 2	5	1	4	3

77

15+ 3	2	5	1– 4	3– 1
3+ 2	5	11+ 1	3	4
1	3	4	3– 2	5
9+ 5	4	3	4– 1	1– 2
5+ 4	1	2 2	5	3

78

2– 1	3	7+ 2	5	1– 4
12+ 5	5+ 1	4	7+ 2	3
4	5 5	3	1	11+ 2
3	15+ 2	1	4	5
2 2	4	5	3	1

79

9+ 4	5	7+ 2	4– 1	1– 3
15+ 3	4	1	5	2
2	9+ 1	4	2– 3	5
1	3	5	9+ 2	3– 4
5	2 2	3	4	1

80

11+ 3	4	15+ 1	2	5
4	5	2	12+ 3	8+ 1
1 1	1– 2	3	5	4
3– 2	4– 1	5	4	3
5	7+ 3	4	3+ 1	2

81

12+ 4	6+ 1	2	3	5 5
5	3	7+ 4	1	14+ 2
14+ 1	10+ 4	5	2	3
3	2 2	1	5	4
2	5	3	3– 4	1

82

16+ 1	5	2	3	4 4
5	7+ 3	4	1 1	15+ 2
2– 4	2	1 1	3– 5	3
12+ 3	4	5	2	1
2 2	2– 1	3	4	5

83

5+ 4	1	11+ 2	6+ 3	5 5
9+ 5	3	4	1	1– 2
1	12+ 4	5	2	3
1– 3	2	1	5	3– 4
2	12+ 5	3	4	1

84

1– 3	4	8+ 2	5	1
1– 4	4– 5	1	13+ 3	2– 2
5	1– 2	12+ 3	1	4
3+ 1	3	4	2	5 5
2	1 1	5	4	3

85

7+ 4	3	16+ 5	2	1
1– 2	1	12+ 4	5	3
3 3	11+ 4	2	1	5
4– 1	5	10+ 3	4	6+ 2
5	2	1 1	3	4

86

9+ 4	5	7+ 2	9+ 1	3
3 3	4	1	5	14+ 2
1– 2	1	4	3	5
4– 1	8+ 3	11+ 5	2	3– 4
5	2	3	4	1

87

8+ 2	3	9+ 4	5	5+ 1
3	2 2	8+ 5	2– 1	4
3– 4	1	2	3	3– 5
1	8+ 5	3	10+ 4	2
1– 5	4	1	2	3

88

15× 3	1	12× 2	40× 5	4
5	12× 4	3	1	2
8× 1	3	60× 4	2	5 5
4	2	5	3	12× 1
10× 2	5	1	4	3

89

10× 1	12× 3	4	30× 2	5
2	5	1	3	24× 4
8× 4	2	15× 5	1	3
15× 5	1	3	20× 4	2
3	8× 4	2	5	1

90

2 2	12× 3	40× 4	5	1
1	4	30× 5	2	24× 3
20× 5	2	3	1	4
4	30× 5	1	60× 3	2
3	1	2	4	5

91

30× 3	2	240× 4	5	1
8× 2	5	1	3 3	4
4	1	5 5	80× 2	3
20× 1	3 3	6× 2	4	5
5	4	3	1	2

92

20× 4	1	120× 2	15× 3	5
5	3	1	4	24× 2
10× 2	5	3 3	1	4
120× 1	2	4	5	3
3	4 4	5	2× 2	1

93

120× 3	4	1	2	5 5
5	10× 1	24× 2	12× 4	3
2	5	4	15× 3	1
2× 1	2	3	5	120× 4
4 4	3	5	1	2

94

240× 3	4	3× 1	10× 2	5
5 5	2	3	48× 4	1
2	5	40× 4	1	3
1	12× 3	2	5 5	4
4	1	5	6× 3	2

95

(18) 3	(9) 4	5	(1) 1	(2) 2
2	3	(9) 4	5	1
(7) 4	1	2	(8) 3	5
(11) 1	5	(1) 3	2	(1) 4
5	(2) 2	(3) 1	4	3

96

(2) 2	(2) 1	(20) 4	5	(7) 3
(9) 5	3	(75) 1	(16) 2	4
4	5	3	1	2
(4) 3	(2) 2	5	4	(4) 1
1	4	(1) 2	3	5

97

(1) 1	(9) 4	5	(1) 2	3
(48) 4	2	(45) 3	(15) 5	1
2	5	1	3	4
3	(7) 1	2	4	5
(12) 5	3	4	(2) 1	2

98

(2) 1	(9) 5	4	(2) 2	(2) 3
2	(10) 3	(50) 5	4	1
(720) 3	1	2	5	(1) 4
4	2	1	3	5
5	4	3	(2) 1	2

99

(8) 4	1	(10) 2	5	(3) 3
2	(7) 4	(9) 5	(3) 3	(2) 1
(90) 5	3	4	(2) 1	2
3	(4) 5	1	2	(13) 4
1	2	3	4	5

100

(5) 5	(40) 4	1	(90) 3	2
(12) 4	(1) 1	2	5	3
1	2	(48) 3	4	5
(8) 3	5	4	(2) 2	1
2	3	(4) 5	1	(4) 4

101

(2) 4	2	(6) 3	(4) 1	5
(45) 5	1	2	(96) 3	4
1	3	(5) 5	4	(3) 2
3	(14) 5	4	2	1
(2) 2	4	1	(8) 5	3

102

(2) 2	(5) 5	(1) 4	3	(2) 1
4	(9) 3	5	1	2
(4) 5	1	(24) 2	4	3
1	(5) 2	3	(100) 5	4
(7) 3	4	(2) 1	2	5

103

[2] 2	[4] 1	5	[36] 4	3
4	[11] 2	[6] 1	3	[5] 5
[3] 3	4	2	[4] 5	1
[6] 1	5	3	[9] 2	4
5	[12] 3	4	1	2

104

[2] 2	[1] 1	[720] 3	4	5
1	[3] 2	[8] 4	[5] 5	3
[20] 3	5	2	1	4
5	4	1	[6] 3	2
[4] 4	3	5	[1] 2	1

105

[2] 4	2	[4] 5	[1] 1	[30] 3
[6] 3	[12] 4	1	5	2
1	3	[40] 2	4	5
2	5	[1] 4	[6] 3	1
[4] 5	1	3	2	[4] 4

106

[4] 5	1	[2] 2	[12] 4	3
[6] 3	[2] 4	1	[8] 5	[11] 2
1	2	[4] 4	3	5
2	[12] 3	5	1	4
[20] 4	5	3	[2] 2	1

107

[15] 3	1	[2] 2	[13] 5	4
5	[12] 3	1	4	[6] 2
4	5	[1] 3	2	1
[7] 2	4	[4] 5	1	3
1	[24] 2	4	3	[5] 5

108

[12] 3	[4] 1	5	[2] 4	2
4	[9] 2	3	[5] 5	[9] 1
[1] 2	4	[5] 1	3	5
1	[13] 5	4	[1] 2	3
5	3	[7] 2	1	4

109

[10] 1	5	[12] 3	4	[2] 2
4	[2] 2	[15] 5	3	[4] 1
[1] 2	[5] 4	[4] 1	5	3
3	1	[2] 4	2	[10] 5
[5] 5	[1] 3	2	1	4

110

[2] 4	[5] 1	[2] 2	[75] 5	[2] 3
2	4	5	3	1
[9] 5	[1] 3	4	[1] 1	[11] 2
3	[8] 5	1	2	4
1	[24] 2	3	4	5

111

(72) 4	2	3	(1) 5	(6) 1
(6) 1	3	(40) 2	4	5
3	(1) 1	5	(2) 2	4
2	(15) 5	4	(2) 1	(1) 3
5	4	1	3	2

112

(20) 5	1	(2) 3	(2) 4	2
1	4	5	(4) 2	(2) 3
(7) 4	3	2	1	5
(1) 2	(4) 5	1	(60) 3	4
3	(2) 2	4	5	(1) 1

113

(1) 3	2	(1) 4	(50) 5	(1) 1
(3) 4	1	3	2	5
(16) 5	3	1	4	2
1	(15) 5	2	3	(12) 4
(2) 2	4	5	(1) 1	3

114

(9) 1	3	(4) 5	(8) 4	2
3	(5) 5	1	2	(19) 4
(2) 2	1	(120) 4	5	3
(13) 5	4	2	3	1
4	2	3	1	5

115

(3+) 1	2	(8×) 4	(160×) 5	8
(7+) 5	1	2	(8) 8	4
2	(12+) 4	(3−) 8	(10×) 1	5
(160×) 4	8	5	2	(2÷) 1
8	5	(4÷) 1	4	2

116

(8) 8	(3+) 2	1	(20×) 4	5
(4−) 5	(2÷) 4	(9+) 8	1	(64×) 2
1	8	(20×) 5	2	4
(8×) 4	1	2	(5) 5	8
2	(1−) 5	4	(7−) 8	1

117

(9+) 4	5	(7−) 8	1	(10×) 2
(4÷) 8	2	(640×) 5	4	1
(4−) 1	8	4	(12+) 2	5
5	1	2	8	(160×) 4
(7+) 2	4	1	5	8

118

(3÷) 9	3	(810×) 6	(6+) 1	5
(4−) 5	1	3	(270×) 9	(5−) 6
(10+) 3	9	5	6	1
1	(15+) 6	9	5	(81×) 3
6	(4−) 5	1	3	9

119

90× 5	3÷ 9	3	5− 6	1
6	4− 1	29+ 5	9	3÷ 3
3	5	6	4− 1	9
8− 1	2÷ 3	9	5	1− 6
9	6	2− 1	3	5

120

45× 5	9	18× 3	1	6
7+ 1	6	2− 5	3	21+ 9
11+ 6	5	10+ 1	9	3
3÷ 9	3	6	1− 5	6+ 1
3	10+ 1	9	6	5

121

2 2	19+ 3	315× 9	7	5
9	7	15+ 3	3− 5	2
15+ 3	5	7	54× 2	9
7	18× 2	3− 5	3÷ 9	3
5	9	2	3	7 7

122

441× 9	7	3− 5	2	3 3
7	3− 5	2	3÷ 3	9
17+ 3	9	378× 7	3− 5	2
5	2	3	28+ 9	7
1− 2	3	9	7	5

123

945× 5	9	3	7	2 2
4− 7	2 2	2− 5	16+ 9	3÷ 3
3	8+ 5	7	2	9
11+ 2	3	9 9	5	35× 7
9	12+ 7	2	3	5

124

11+ 5	6	2÷ 4	8	1− 7
336× 8	560× 5	7	4	6
6	4	27+ 8	175× 7	5
7	8	6	5	192× 4
11+ 4	7	5	6	8

125

210× 5	6	7	2÷ 8	4
1344× 6	100× 5	192× 8	4	7 7
7	4	5	6	13+ 8
4	8	1− 6	7	5
224× 8	7	4	1− 5	6

126

9+ 5	192× 8	6	4	1− 7
4	210× 7	5	6	8
1− 7	6	2÷ 8	20× 5	4
192× 8	5 5	4	210× 7	6
6	4	1− 7	8	5

127

3÷ 9	15× 5	3	1	7− 8
3	22+ 8	5	9	1
4− 5	1	72× 8	3÷ 3	9
7− 1	3÷ 3	9	18+ 8	5
8	9	1	5	3 3

128

675× 5	3	3÷ 9	14+ 8	1
8− 9	1	3	5	11+ 8
1	5	216× 8	9	3
5− 8	9	1	3	14+ 5
3	14+ 8	5	1	9

129

45× 9	14+ 1	8	5	24× 3
5	13+ 9	1	3	8
15× 1	3	3÷ 9	3240× 8	5
5− 8	5	3	9	1 1
3	3− 8	5	1	9

130

21+ 8	2− 6	4	7 7	3÷ 2
7	3584× 2	8	4	6
6	8	11+ 7	2	4 4
4 4	7	2	2688× 6	8
2− 2	4	6 6	8	7

131

32× 2	1− 7	6	4− 4	8
8	2	25+ 7	12+ 6	4
672× 7	6	4	4÷ 8	2
6	4	8	2	294× 7
4	6− 8	2	7	6

132

18+ 6	4	8	5− 2	7
2÷ 4	1− 6	7	3072× 8	2
2	112× 8	3− 4	7	6
7	2	6 6	4	8
112× 8	7	2	2− 6	4

133

378× 7	6	9 9	1− 3	2
9	2 2	2÷ 3	6	4− 7
4536× 6	23+ 9	7	36× 2	3
3	7	2	9	15+ 6
2	3	6	7	9

134

6 6	3÷ 9	3	5− 7	2
21× 7	2 2	3− 6	9	37+ 3
3	2÷ 6	5− 7	2	9
7− 2	3	9 9	6	7
9	7	2	3	6 6

135

9+ 4	3	2 2	5− 1	6	10+ 5
2	4− 6	1− 4	11+ 5	2− 3	1
8+ 3	2	5	6	1	4
5	6+ 1	3− 3	12+ 2	4	6
1	4	6	8+ 3	3− 5	2
11+ 6	5	1	4	5+ 2	3

136

6 6	12+ 2	1− 3	11+ 5	3− 4	1
2	5	4	6	2− 1	3
3	15+ 6	5	1 1	15+ 2	4
1	3	2 2	4	5	11+ 6
3− 4	1	5− 6	5+ 2	3− 3	5
9+ 5	4	1	3	6	2 2

137

5+ 3	2	2− 1	1− 5	4	11+ 6
6+ 2	4	3	5− 1	6	5
5− 1	1− 5	6	10+ 3	2	3− 4
6	1− 3	2	4 4	5	1
1− 5	10+ 6	4	4− 2	2− 1	3
4	4− 1	5	6	5+ 3	2

138

8+ 1	11+ 5	1− 3	2− 4	6	2 2
3	6	4	9+ 2	5	4+ 1
4	4− 1	5	6 6	2	3
1− 5	4	2 2	4− 1	9+ 3	6
13+ 2	3	5− 6	5	3− 1	4
6	2	1	12+ 3	4	5

139

15+ 5	1	6+ 2	4	3− 3	6
4	5	5− 1	6	5+ 2	3
10+ 1	5+ 2	3	11+ 5	6	9+ 4
3	6	1− 5	1− 2	4	1
12+ 6	3 3	4	1	7+ 5	2
2	4	3− 6	3	4− 1	5

140

13+ 2	1	3	1− 4	5	6 6
5− 1	6	5	2	3 3	9+ 4
1− 3	3+ 2	1	6 6	12+ 4	5
4	13+ 5	6	8+ 3	2	1
14+ 6	4 4	2	5	5− 1	3
5	3	5+ 4	1	6	2

141

(3) 3	(23+) 4	6	(3+) 1	2	(13+) 5
(12+) 1	(5) 5	3	4	6	2
2	3	(5) 5	6	(3–) 1	4
(15+) 6	2	4	(5) 5	(9+) 3	(2–) 1
5	(5–) 6	(1–) 1	2	4	3
4	1	2	(8+) 3	5	(6) 6

142

(19+) 5	6	1	3	4	(2) 2
(5+) 2	3	(1–) 4	5	(1) 1	(19+) 6
(5–) 1	(8+) 4	2	(14+) 6	3	5
6	2	(7+) 3	4	5	1
(10+) 4	5	(5–) 6	1	(4–) 2	3
(3) 3	1	(3–) 5	2	6	4

143

(5–) 6	(7+) 3	4	(19+) 1	5	(2) 2
1	(7+) 2	5	(11+) 6	4	3
(8+) 3	(5+) 4	1	5	(2) 2	6
5	(5–) 1	6	(2–) 2	(3–) 3	(5+) 4
(13+) 2	5	(3) 3	4	6	1
(4) 4	6	(1–) 2	3	(4–) 1	5

144

(3+) 2	1	(2–) 5	(7+) 3	(17+) 6	(4) 4
(11+) 1	(2) 2	3	4	5	6
4	(12+) 3	2	(6) 6	(6+) 1	5
6	(19+) 4	1	5	(1–) 2	3
(14+) 3	5	6	1	(7+) 4	2
5	6	4	(1–) 2	3	1

145

(2–) 2	(3–) 1	4	(14+) 6	3	5
4	(3–) 5	(3–) 6	(6+) 1	2	3
(1) 1	2	3	(1–) 4	5	(5–) 6
(3–) 3	(12+) 6	2	(13+) 5	(4) 4	1
6	4	5	3	(1–) 1	2
(9+) 5	3	1	(8+) 2	6	(4) 4

146

(2–) 3	1	(10+) 4	6	(13+) 2	(5) 5
(5–) 6	(9+) 4	5	2	1	3
1	(20+) 2	(12+) 3	4	5	(5–) 6
(1–) 5	6	2	3	(7+) 4	1
4	5	6	1	3	(2–) 2
(5+) 2	3	(1) 1	(1–) 5	6	4

147

2– 4	2	9+ 3	15+ 6	5	6+ 1
6 6	1	2	3	4	5
6+ 2	1– 4	5	10+ 1	6	3
1	1– 6	1– 4	5	5+ 3	12+ 2
3	5	15+ 1	7+ 4	2	6
5	3	6	2	1	4

148

13+ 4	5	4+ 1	3	8+ 2	6
1– 1	4	25+ 6	7+ 2	2– 3	5
2	2– 3	4	5	5– 6	1
1– 5	1	3	6	4	2
6	2 2	1– 5	4	8+ 1	3
9+ 3	6	2 2	6+ 1	5	4

149

22+ 2	1	6	5	4	3 3
4	10+ 3	6+ 1	15+ 6	2 2	23+ 5
5– 1	2	5	3	6	4
6	5	2 2	7+ 4	3	1
15+ 5	4	6+ 3	2	1	6
3 3	6	5+ 4	1	5	2

150

4 4	23+ 1	5	6	2	3
23+ 2	9+ 4	3	6+ 5	1	6
1	2	4 4	3– 3	6	4– 5
6	8+ 5	4– 2	8+ 4	3	1
5	3	6	1	4 4	11+ 2
3	6	3+ 1	2	5	4

151

11+ 5	4– 6	2	2– 1	3	4 4
4	2	2– 3	8+ 6	6+ 1	5
5– 1	11+ 4	5	2	13+ 6	2– 3
6	3	4	5	2	1
2– 3	1	5– 6	20+ 4	5	2
3– 2	5	1	3	4 4	6

152

18+ 2	1	1– 4	3	1– 6	5
1	3– 2	5	9+ 6	3	9+ 4
3	5– 6	1	1– 5	4	2
6	5	3 3	7+ 4	2	1
13+ 5	4	8+ 6	2	4– 1	9+ 3
4	3 3	2	1	5	6

153

12+ 2	4	6	9+ 5	1	3
14+ 1	6	2	11+ 3	5	5+ 4
1– 4	2 2	5	4– 6	3	1
3	9+ 1	4	2	12+ 6	11+ 5
5 5	3	1	4	2	6
14+ 6	5	3	7+ 1	4	2

154

20+ 5	4	6	1– 2	3	1 1
4	19+ 6	1	13+ 3	5	4– 2
1– 2	3	5	1	4	6
3	5	2 2	19+ 6	15+ 1	4
8+ 1	2 2	3	4	6	5
6	1	1– 4	5	2	3

155

360× 6	3	5	8× 2	1	4
4	180× 5	6	1	6× 2	3
40× 1	2	3	432× 4	6	5 5
2	4	24× 1	5 5	3	6
5	18× 6	2	3	4 4	1
3	1	4	60× 6	5	2

156

40× 2	144× 1	4	18× 6	3	30× 5
4	5	6	80× 1	2	3
18× 1	2	3	4	5	6 6
3	6	10× 2	5	4	1
90× 6	20× 4	5	6× 3	1	48× 2
5	3	1	2	6	4

157

90× 5	2	3	144× 1	4	6
12× 1	3	8× 2	4	6	10× 5
3	4	30× 5	6	1	2
24× 4	6	6× 1	120× 2	15× 5	3
12× 2	1	6	5	3	4
6	60× 5	4	3	2 2	1

158

12× 2	1	20× 5	4	6 6	12× 3
5 5	6	72× 3	2	4	1
18× 3	4	1	12× 6	2	6480× 5
6	3	16× 2	25× 1	5	4
20× 1	2	4	5	3	6
4	5	6	3	1	2 2

159

24× 1	6	20× 4	5	36× 3	2
60× 5	4	6× 3	2	6	1
3	1	30× 5	6	48× 2	4
4	720× 5	2	3× 3	1	6
6× 2	3	6	1	4 4	15× 5
6	2	1	20× 4	5	3

160

20× 4	12× 1	3	36× 6	2	30× 5
5	4	30× 2	1	3	6
12× 1	2	5	3	720× 6	4
6	20× 5	4	2× 2	1	3
18× 3	6	20× 1	5	4 4	2
6× 2	3	6 6	4	5	1

161

1 1	4320× 3	4	30× 5	6	2
17280× 4	1 1	2	3	5	6
3	4	180× 5	6	2× 2	1
10× 5	2	6	12× 1	3	20× 4
2	6	3 3	4	1	5
6	5	8× 1	2	4	3 3

162

11 5	6	2 2	15 1	4	3
5 6	80 4	1	2	54 3	5
1	5	4	3	6	2 2
1 4	19 3	6	5	2 2	5 1
3	8 2	5	21 6	1	4
2	1	3	4	5	6

163

100 5	12 2	4	6	36 1	3
4	5	5 6	1	3 3	2
3 2	30 3	5	1 4	6	1
1	4 4	2	3	22 5	6
108 3	6	7 1	2	4	5
6	3 1	3	5 5	2	4

164

11 4	1 5	6	12 2	3 3	24 1
5	5 6	1	3	2	4
2	8 3	6 5	1	96 4	6
3 1	2	3	4	6	50 5
3	1	2 4	30 6	5	2
24 6	4	2	5	3 1	3

165

6: 1	3	800: 4	5	2	90: 6
2	5: 1	6	4: 4	5	3
1080: 3	3: 2	12: 1	6	4	5
4	6	5	6: 1	9: 3	2
6	5	3	2	5: 1	4
11: 5	4	2	3	6	1: 1

166

320: 5	2	4	10: 6	1	3
4	72: 3	6	1: 1	2	11: 5
2	4	10: 1	5	3: 3	6
10: 1	11: 5	2	36: 3	6	4: 4
3	6	5: 5	2	12: 4	1
6	3: 1	3	4: 4	5	2

167

3: 1	4	180: 2	5	6	3
11: 5	6	32: 4	2	15: 3	1
18: 3	1	11: 6	4	5	2: 2
6	2: 3	5	3: 1	2: 2	4
16: 2	5	1	3	4	1: 6
4	2	2: 3	6	1: 1	5

168

40: 4	11: 6	5	6: 1	3	2
2	5	1	13: 4	6	3
12: 3	4	11: 6	5	2: 2	1
90: 5	3	16: 2	90: 6	24: 1	4
1	2	4	3	5	6
6	1	1: 3	2	1: 4	5

169

9: 4	1	3: 2	6	75: 5	3
5: 6	4	6: 1	3	2	5
1	900: 2	3	5	13: 4	6
5	6	32: 4	2	3	1: 1
1: 3	11: 5	6	4	7: 1	2
2	2: 3	5	5: 1	6	4

170

3: 6	40: 2	4	5	3: 3	1
2	2: 4	5: 3	10: 6	1	100: 5
1: 1	6	2	3	5	4
180: 4	3	5: 6	1	10: 4	36: 2
4	25: 5	1	16: 2	6	3
3	1	5	4	2	6

171

(5) 6	(1) 1	(80) 4	5	(13) 2	3
1	(3) 6	2	4	3	5
(2) 2	(144) 4	1	3	(60) 5	6
(60) 4	5	(3) 3	6	(3) 1	2
(30) 5	3	(30) 6	2	4	(1) 1
3	2	5	(1) 1	(24) 6	4

172

(8) 2	(5) 1	6	(60) 4	3	5
4	(40) 5	(16) 3	1	2	6
1	2	(15) 4	6	5	3
(5) 5	4	(7) 2	3	(24) 6	1
(108) 6	(2) 3	5	2	1	(2) 4
3	6	(6) 1	5	4	2

173

(30) 5	2	3	(15) 1	4	6
(3) 3	(2) 5	(4) 6	(8) 2	1	4
1	3	2	4	(1) 6	5
(18) 2	(16) 1	4	(90) 6	5	(3) 3
6	4	(4) 5	3	(2) 2	(6) 1
4	6	1	(5) 5	3	2

174

(2) 4	6	(1) 1	(30) 5	(1) 3	2
(15) 5	4	6	2	(3) 1	3
(6) 2	(40) 5	4	3	(1200) 6	1
1	2	(1) 3	4	5	(1) 6
3	(10) 1	(3) 2	6	4	5
6	3	5	1	2	(4) 4

175

(2) 1	2	(60) 5	(13) 3	4	6
(1) 4	1	6	2	(13) 5	3
3	(7) 4	(2) 2	(5) 6	1	5
(3) 2	3	1	(15) 5	6	(4) 4
5	(17) 6	(14) 3	4	(2) 2	1
6	5	4	1	3	2

176

(1) 2	3	(6) 6	(32) 4	(4) 1	5
(2) 5	(5) 6	1	2	4	(13) 3
3	(3) 2	5	(5) 1	6	4
(1152) 4	(10) 1	2	(45) 3	5	6
1	5	4	(13) 6	3	(2) 2
6	4	3	5	2	1

177

[2] 2	4	[3] 3	[800] 5	[18] 1	6
[90] 5	[5] 6	2	4	3	[2] 1
3	1	5	[10] 6	4	2
6	[3] 3	1	[3] 2	5	[1] 4
[8] 1	2	4	[2] 3	6	5
[4] 4	5	[7] 6	1	[1] 2	3

178

[288] 4	3	2	1	[14] 5	[6] 6
6	[7] 5	1	2	4	3
2	1	[1] 4	[15] 6	[3] 3	[1] 5
1	[144] 2	3	5	[5] 6	4
3	6	[5] 5	4	1	[4] 2
[5] 5	4	[2] 6	3	2	1

179

[144] 4	6	[6] 1	5	[6] 2	3
6	[2] 2	[1] 4	3	1	[22] 5
[2] 1	[24] 3	2	[120] 4	[90] 5	6
2	4	5	6	3	1
[6] 5	1	[6] 3	2	6	4
[14] 3	5	6	1	4	2

180

[2700] 6	[3] 1	4	5	3	[2] 2
1	3	5	2	[4] 4	[7200] 6
[2] 2	[11] 6	[3] 1	3	5	4
[40] 4	5	[13] 6	1	2	[3] 3
5	2	3	4	6	1
[144] 3	4	2	6	1	5

181

[5] 5	[2] 6	3	[80] 1	4	2
[5] 6	[3] 3	2	5	[4] 1	[10] 4
1	[960] 2	[72] 4	3	5	6
2	1	6	[4] 4	[90] 3	5
4	[4] 5	1	[72] 6	2	3
3	4	5	2	6	[1] 1

182

[14+] 9	5	[5−] 1	6	[8×] 4	2
[5−] 4	9	[6] 6	[540×] 2	5	1
[3÷] 2	[100×] 1	5	9	6	[4] 4
6	[72×] 2	4	1	[2430×] 9	5
[30×] 1	4	9	5	[2÷] 2	6
5	6	[2÷] 2	4	1	9

183

432× 9	4	1	6	2	30× 5
11+ 4	5	18× 2	9	30× 1	6
3+ 1	2	120× 4	5	6	15+ 9
2	8– 9	6	2÷ 4	5	1
11+ 6	1	5	2	13+ 9	2÷ 4
5	16+ 6	9	1	4	2

184

180× 9	20× 4	5	1	3÷ 6	2
1	22+ 2	6	14+ 9	5	13+ 4
4	5	1	4– 6	2	9
6× 6	1	9	3– 2	1– 4	5
7+ 2	3– 6	4	5	8– 9	1
5	9	2÷ 2	4	5– 1	6

185

15× 5	105× 3	1	7	14+ 6	8
3	6 6	20+ 5	1	1– 8	7
17+ 1	7	8	5	3 3	5– 6
8	5	3	1– 6	7	1
13+ 7	7– 8	6 6	3÷ 3	1	13+ 5
6	1	1– 7	8	5	3

186

13+ 6	7	4– 1	5	5– 3	8
1 1	48× 6	8	2– 3	5	630× 7
840× 3	30× 5	6	15+ 7	8	1
7	13+ 1	5	20+ 8	6	3
8	5– 3	7	6	6– 1	5
5	8	3÷ 3	1	7	6

187

35× 7	3÷ 3	1	14+ 8	6	5 5
1	5	24× 3	5– 6	1– 8	7
30× 5	2– 6	8	1	4– 7	720× 3
6	8	1– 5	12+ 7	3	1
19+ 3	7	6	5	1 1	8
8	1	4– 7	3	5	6

188

1– 8	9	6480× 5	3÷ 6	2	3 3
11+ 3	6	9	10× 2	30+ 5	8
2	3	6	5	8	9
5 5	8	3÷ 3	9	3÷ 6	2
15+ 9	3– 5	2	5– 8	3	30× 6
6	4÷ 2	8	3÷ 3	9	5

189

3÷ 9	11+ 6	48× 8	5− 3	2700× 2	5
3	5	6	8	9	4÷ 2
6− 2	3÷ 9	1− 3	5	6	8
8	3	2	28+ 9	5	54× 6
30× 6	4÷ 8	4− 5	3÷ 2	3	9
5	2	9	6	8	3

190

9 9	3÷ 2	6	360× 8	5	36× 3
2− 3	19+ 5	8	9	6	2
5	6	8+ 2	3	6480× 9	3− 8
108× 6	9	3	2 2	8	5
2	19+ 8	5	6	3	54× 9
8	3	45× 9	5	2 2	6

191

1− 8	3 3	75× 5	31+ 6	7	9
9	5	3	21+ 7	8	6
35× 5	7	168× 8	9 9	6	3
14+ 6	8	7	3	14+ 9	2− 5
2÷ 3	6	3− 9	3− 8	5	7
2− 7	9	6	5	11+ 3	8

192

2÷ 6	3	11+ 8	3− 5	3− 9	27216× 7
2− 5	7	3	8	6	9
11+ 3	4− 5	13+ 7	9	8	6
8	9	6	28+ 7	5	3
3− 9	6	45× 5	2÷ 3	21× 7	8
1− 7	8	9	6	3	5

193

7560× 3	8	5	22+ 6	7	9
3− 8	4− 3	7	9	14+ 5	6
5	7	17+ 9	24+ 8	6	3
2− 9	1− 6	8	7	3	40× 5
7	5	972× 6	3÷ 3	9	8
6	9	3	280× 5	8	7

194

3÷ 3	1	40× 8	720× 6	2− 5	7
21+ 6	7	5	1	3	8
1	14+ 3	21+ 6	8	7	5
7	6	10+ 3	2− 5	9+ 8	1
40× 8	5	7	3	1	2÷ 6
5	9+ 8	1	42× 7	6	3

195

240× 5	6	17640× 7	3÷ 3	7– 1	8
8	5	3	1	13+ 6	7
5– 1	3	200× 5	15+ 8	7	6 6
6	7	8	5	3÷ 3	1
21× 7	8	5– 1	6	5 5	16+ 3
3	1	42× 6	7	8	5

196

48× 8	5– 1	6	2– 7	5	4– 3
6	13+ 8	5	24+ 3	1	7
3÷ 1	3	15+ 7	6+ 5	8	6
2÷ 3	6	8	1	21+ 7	5
525× 7	5	3	8	6	1
5	7 7	5– 1	6	11+ 3	8

197

2 2	432× 8	1– 6	7	5– 9	4
6	9	17+ 7	72× 4	165888× 8	3÷ 2
1– 7	2	8	9	4	6
8	3– 7	4	2	6	9 9
13+ 4	6	9	8	2	22+ 7
9	2÷ 4	2	6 6	7	8

198

3÷ 2	6	21+ 4	2– 9	7	8 8
9 9	5– 2	8	96× 6	112× 4	7
290304× 6	7	9	8	2	4
8	13+ 4	2	22+ 7	6	9
4	8 8	7	7– 2	9	3÷ 6
7	9	6	4	8 8	2

199

112896× 8	7	9	4	3÷ 2	6
4– 6	288× 4	1– 7	8	7– 9	2
2	9	6	7	17+ 4	8 8
7 7	8	38+ 4	3÷ 2	6	36× 9
7– 9	2	8	6	7	4
2– 4	6	2	9	8	7

200

10+ 1	30× 5	6	2÷ 8	5– 9	4
9	30× 1	1– 5	4	14+ 6	8
5	6	4	2160× 9	8	8– 1
288× 4	8	7– 1	6	5	9
14+ 6	9	8	6+ 1	9+ 4	5
8	5– 4	9	5	5– 1	6

201

1 1	8– 9	120× 4	6	22+ 5	8
3– 8	1	5	288× 4	6 6	9
5	2– 4	6	8	9	13+ 1
15552× 9	15+ 6	1	4– 5	8	4
4	5 5	8	9	5– 1	6
6	8	9	4÷ 1	4	5 5

202

9+ 3	2	4	14+ 5	8+ 7	1	6 6
11+ 5	17+ 6	2	3	4	6– 7	1
6	4	7	3+ 1	2	7+ 3	2– 5
3+ 1	10+ 3	5	2	21+ 6	4	7
2	13+ 1	3	7	5	2– 6	4
3– 4	7	5– 1	6	2– 3	5	1– 2
7	5	2– 6	4	3+ 1	2	3

203

12+ 7	4	3+ 1	2	17+ 6	8+ 5	6+ 3
1	18+ 5	10+ 6	4	7	3	2
2– 5	7	4	6+ 3	2	12+ 6	1
3	6	12+ 7	5	1	2	4
4– 6	5+ 3	2	4– 1	5	3– 4	7
2	4+ 1	3	10+ 6	4	7 7	1– 5
4 4	7+ 2	5	11+ 7	3	1	6

204

1– 1	2	14+ 4	3	7	11+ 6	5
25+ 6	4– 5	1	13+ 2	3	4	6– 7
3	4– 6	2	4	15+ 5	7	1
4	15+ 3	5	7	1	6+ 2	6 6
5	6– 1	7	17+ 6	2	3	14+ 4
7	7+ 4	3	5	6	1	2
5– 2	7	11+ 6	1	4	5	3

205

3+ 2	1	17+ 5	6	3	3– 4	7
18+ 7	6	5+ 2	3	12+ 4	5	1
2– 1	5	3	5– 7	2	6 6	1– 4
3	3– 4	7	2	5– 6	1	5
18+ 5	7	6	3– 4	6+ 1	2	3
10+ 6	1– 3	5+ 4	1	12+ 5	10+ 7	8+ 2
4	2	1	5 5	7	3	6

206

11+ 5	8+ 3	4	13+ 7	6	7+ 2	6– 1
6	1	10+ 2	3	4	5	7
9+ 3	4	19+ 5	1	6– 7	15+ 6	2
2	5	3	6	1	7	12+ 4
6– 1	15+ 6	7	6+ 2	10+ 5	4	3
7	2	14+ 6	4	3 3	1	5
11+ 4	7	1	5	2	9+ 3	6

207

2− 3	4− 2	19+ 4	5	7	1	6 6
5	6	2	1− 3	4	6− 7	1
17+ 6	4	7	9+ 1	2 2	8+ 3	5
3+ 1	2− 3	5	2	6	15+ 4	7
2	6− 1	2− 3	13+ 7	11+ 5	6	4
3− 4	7	1	6	2− 3	5	1− 2
7	15+ 5	6	4	3+ 1	2	3

208

11+ 7	2− 3	28+ 6	2	1 1	1− 4	5
4	1	2− 3	7	7+ 2	5	6 6
12+ 5	7	1	3	4	6	5− 2
6+ 1	2− 2	4	5 5	3− 6	3	7
2	27+ 4	5	6	2− 3	6− 7	1
3	13+ 6	7	1	5	3+ 2	7+ 4
11+ 6	5	2 2	4	7	1	3

209

1− 7	5− 6	1	9+ 4	3 3	7+ 2	5
6	6− 1	17+ 2	5	3− 7	4	3− 3
4	7	5	6+ 1	2	3	6
1	2	3	18+ 6	28+ 4	5	7
1− 3	4	6	2	5	6− 7	1
2	2− 5	7	10+ 3	6	1	2− 4
8+ 5	3	4 4	7	7+ 1	6	2

210

1− 3	4	4− 1	5	34+ 7	6	2 2
9+ 2	6	7	3	5	1 1	19+ 4
7	9+ 3	6+ 4	2	11+ 6	5	1
4	2	11+ 6	6− 7	1	3	2− 5
19+ 6	7	5	1	4	2	3
5	1	3 3	12+ 6	2	4	1− 7
1 1	3− 5	2	4	10+ 3	7	6

211

13+ 4	1	5	3 3	19+ 7	6	2
3+ 2	3	6− 7	1	20+ 6	4	2− 5
1	10+ 2	6	11+ 5	4	7	3
5 5	13+ 7	2	6	3	2− 1	12+ 4
13+ 7	6	11+ 4	2	5	3	1
6	15+ 5	3	4	1	2	7
1− 3	4	1 1	5− 7	2	11+ 5	6

212

1− 5	24+ 1	2	5+ 3	17+ 4	7	6
6	12+ 4	1	2	3 3	2− 5	7
6+ 2	3	6	3− 4	7	7+ 1	6+ 5
3	5	7	1− 6	2	4	1
1	8+ 6	3	7	17+ 5	2− 2	4
3− 7	2	4	5	1	6	5+ 3
4	7 7	4− 5	1	9+ 6	3	2

213

29+ 4	7	4– 5	1	13+ 6	3	6+ 2
6 6	5	5– 7	2	7+ 1	4	3
3– 5	6	4– 3	7	4	2	1
2	1	6	3 3	5– 7	12+ 5	1– 4
6– 1	13+ 3	4	28+ 6	2	7	5
7	4	2	5	3	1	6
6+ 3	2	1	15+ 4	5	6	7

214

12+ 2	4	5– 6	1	8+ 5	3	2– 7
6	7 7	16+ 4	9+ 3	21+ 1	2	5
3– 1	5	7	4	2	6	3
4	10+ 2	5	7 7	8+ 3	1	6
13+ 7	6	3	4– 2	4	16+ 5	1
5 5	24+ 3	1	6	7	4	2
4+ 3	1	2	5	6	7	4 4

215

16+ 6	3	6+ 2	4	20+ 7	13+ 5	1
3	15+ 2	6+ 5	1	6	7	4
4	7	6	3 3	8+ 5	1	2
6– 1	12+ 5	7	3– 2	3	10+ 4	6
7	6+ 4	1	5	30+ 2	6	3
13+ 5	1	9+ 3	6	4	2 2	7
2	6	12+ 4	7	1	3	5

216

1– 4	5	10+ 7	2	1	19+ 6	3
9+ 5	4– 2	3– 6	6– 1	13+ 4	3	7
1	6	3	7	5	4	3– 2
3	10+ 1	10+ 4	6 6	11+ 7	2	5
6	3	1	5	2	35+ 7	4 4
11+ 7	4	2	3	6	5	5– 1
2 2	7	5	8+ 4	3	1	6

217

18+ 6	2	8+ 3	1	4	5 5	6– 7
4	6	19+ 5	3	5– 7	2	1
5 5	7	1	8+ 2	24+ 3	4	6
6+ 1	3	2	4	6	26+ 7	5
2	4– 1	33+ 4	7	5	6	3 3
3	5	7	6	2	7+ 1	4
3– 7	4	6	5	1	3	2

218

11+ 1	23+ 4	6	7	2	14+ 3	5
2	3	5	3+ 1	4	23+ 7	6
2– 5	7	6– 1	2	16+ 3	6	7+ 4
2– 3	1	7	5 5	6	4	2
4– 6	2	5+ 3	1– 4	7	5	1
17+ 4	6	2	3	12+ 5	1	12+ 7
7	9+ 5	4	6	1	2	3

219

8+ 4	2	18+ 6	10+ 1	5	14+ 3	7
2	7	5	4	19+ 6	1	3
15+ 5	8+ 1	4	3	7	6	5+ 2
3	18+ 6	7	5	7+ 4	2	1
7	9+ 5	3	5− 2	1	15+ 4	6
11+ 6	3 3	1	7	2	5	16+ 4
1	4	11+ 2	6	3	7	5

220

5− 2	7+ 1	14+ 3	7	6 6	1− 4	5
7	5	1	4	9+ 3	1− 6	15+ 2
4+ 1	4− 3	7	2	4	5	6
3	15+ 2	2− 4	6	4− 5	1	7
2− 4	7	6	8+ 5	1	12+ 2	3
6	4 4	7+ 5	11+ 3	2	7	3− 1
11+ 5	6	2	1	7	3 3	4

221

4 4	24+ 1	2	7	14+ 6	3	5
7	3	3− 6	4	3− 5	2	3− 1
6+ 5	9+ 7	3	11+ 6	6+ 2	1 1	4
1	2	4− 5	3	4	1− 6	4− 7
9+ 6	4 4	1	2	6− 7	5	3
3	13+ 6	7	22+ 5	1	4	2
11+ 2	5	4	1	3	7	6 6

222

21× 7	1	240× 2	5	72× 3	6	8× 4
3	70× 5	7	1	1260× 6	4	2
84× 1	2	4	6	7	15× 5	3
4	7	3	2 2	5	1	6
1080× 6	1680× 4	5	7	12× 2	3	1
5	6	1	3	4× 4	2	245× 7
6× 2	3	6	4	1	7	5

223

840× 6	5	7	2× 1	12× 4	3	6× 2
1	84× 7	4	2	15× 5	60× 6	3
12× 4	1	120× 5	6	3	2	7 7
3	2	6	4	42× 7	5	4× 1
30× 5	3	2	105× 7	6	1	4
48× 2	4	3	5	1	168× 7	30× 6
7 7	6	1	3	2	4	5

224

60× 4	3	96× 2	6	210× 7	5	1
5	1	420× 7	2	126× 3	6	20× 4
7× 7	6	1	4	2	3	5
1	2	3 3	140× 5	4	7	126× 6
48× 6	5	360× 4	7	1	2 2	3
2	4	5	3	6	1	7
126× 3	7	6	1	40× 5	4	2

225

28× 7	4	1050× 1	180× 2	6	5	3
1	5	6	48× 4	672× 7	3	2
5	7	4	3	2	6× 6	1
18× 3	42× 6	7	5 5	1	2	4
6	504× 3	2	1	5 5	140× 4	7
8× 2	1	15× 3	6	12× 4	7 7	5
4	2	5	7	3	1	6 6

226

300× 6	140× 7	5	24× 4	1	42× 3	2
5	1	4	2	3	6 6	7
24× 3	5	2	630× 1	42× 6	7	20× 4
2	4	3	6	7	5	1
126× 1	3	7	5	96× 4	60× 2	6
28× 7	6	6× 1	105× 3	2	4	5
4	2 2	6	7	5	1	3

227

30× 1	15120× 3	6	4	70× 2	7	5
5	6	18× 3	7	1	8× 2	4
7 7	48× 4	2	3	6	5	42× 1
3	2	4 4	1	120× 5	6	7
2	7350× 5	7	6	4	36× 1	3
120× 6	7 7	1	5	3	4	12× 2
4	1	5	2	7	3	6

228

14× 7	1	7560× 6	2	16× 4	126× 3	5 5
2	3	5	4	1	7	6
6	7	8× 1	30× 3	2	120× 5	4
120× 1	2	4	7 7	5	6	252× 3
5	4	126× 7	6	3	2	1
60× 4	6	30× 3	5	7	1	56× 2
3	5	2	1	6	4	7

229

13 1	216 3	48 6	4	2	140 7	5
5	6	3	42 7	3 1	2	4
7	4	2	3	6 6	150 5	6 1
1 3	280 2	4	1	5	6	7
2	5	13 7	6	36 4	1	3
24 6	7	1	19 5	3	7 4	3 2
4	1	5	2	7	3	6

230

16 2	3	13 6	7	24 1	140 4	5
11 1	5	4	2	3	7	6 6
3	6	6 1	3 4	2 7	7 5	2
4	2	7	1	5	216 6	3
13 7	1	75 5	3	6	2	336 4
6	7 4	3	5	2 2	1	7
14 5	7	2	6 6	4	3	1

231

(180) 5	6	(168) 7	4	(8) 2	1	(3) 3
6	(72) 3	4	2	5	(2) 7	(6) 1
(6) 2	(2) 4	6	1	3	5	7
1	2	(15) 3	5	7	(4) 4	(30) 6
3	(2) 1	2	(7) 7	(24) 4	6	5
(28) 4	7	(4) 5	(3) 6	(3) 1	3	(2) 2
(35) 7	5	1	3	(4) 6	2	4

232

(105) 7	(3) 3	1	(120) 4	6	5	(7) 2
3	(17) 6	(140) 5	2	(3) 7	4	1
5	7	2	(1890) 6	3	(12) 1	4
(1) 1	4	7	3	2	6	(11) 5
(2) 4	(4) 5	(2) 3	7	(1) 1	2	6
2	1	6	5	(1) 4	3	(147) 7
(3) 6	2	(20) 4	1	5	7	3

233

(2) 5	(84) 7	(2) 4	2	(90) 3	1	6
7	2	6	(6) 1	(96) 4	(3) 3	5
(8) 3	5	(3) 2	7	6	4	1
(6) 2	3	1	(14) 5	(4116) 7	6	(2) 4
1	(360) 6	3	4	5	7	2
(24) 6	4	5	(6) 3	1	2	7
4	1	(13) 7	6	2	(15) 5	3

234

(4) 5	1	(168) 2	3	4	7	(1050) 6
(3) 6	(3) 4	1	(9) 2	3	5	7
2	(270) 3	(13) 6	4	(70) 7	(7) 1	5
(84) 3	5	7	(13) 6	2	4	(1) 1
1	6	3	7	5	2	(15) 4
7	(2) 2	4	(6) 5	1	6	3
4	(630) 7	5	1	6	3	2

235

(56) 2	4	(15) 3	5	(28) 7	1	(2) 6
7	(13) 6	(11) 2	(13) 1	5	4	3
(45) 1	7	6	3	2	5	(56) 4
3	(5) 5	(5) 4	(13) 7	(7) 1	6	2
5	3	1	6	(21) 4	2	7
(48) 4	1	7	2	6	(63) 3	(4) 5
6	2	(1) 5	4	3	7	1

236

(2) 1	2	(20) 4	(3) 7	(15) 6	(2) 5	3
(11) 2	5	1	4	3	6	(35) 7
4	(108) 6	3	2	(6) 1	7	5
5	3	(168) 7	6	4	(2) 2	1
(13) 7	(1680) 4	6	(2) 3	(15) 5	1	2
6	(1) 1	2	5	7	(72) 3	4
(3) 3	7	5	(7) 1	2	4	6

237

(336) 7	4	(20) 2	5	(2) 3	1	(5) 6
4	3	(35) 5	2	(13) 7	6	1
(3) 3	7	1	(24) 4	6	2	5
1	(5) 5	(27) 3	(11) 6	2	(13) 4	7
(3) 6	(7) 1	4	3	(5) 5	7	2
2	6	7	(6) 1	4	(45) 5	3
(3) 5	2	6	7	1	3	(4) 4

238

(3) 3	(3) 4	1	(144) 6	2	(30) 5	(10) 7
(2) 2	(7) 7	(1) 4	5	1	6	3
1	(7) 6	(2) 2	3	4	(18) 7	5
(20) 4	1	3	(70) 7	5	2	6
5	2	6	(6) 1	7	(216) 3	(2) 4
(4) 7	3	(70) 5	(4) 4	6	1	2
(11) 6	5	7	2	3	4	(1) 1

239

(5) 5	(1152) 6	(2) 2	1	(588) 3	7	4
3	2	(16) 1	4	(210) 6	5	7
1	(105) 3	6	5	7	(2) 4	2
4	5	7	(108) 3	2	6	(5) 1
2	4	(15) 5	(6) 7	1	3	6
(1176) 7	1	3	(2) 6	4	(1) 2	(8) 5
6	7	4	(3) 2	5	1	3

240

(24) 6	4	(54) 3	(2) 5	7	(2) 2	1
1	(16) 2	6	3	(2) 5	(3) 7	(20) 4
7	6	(1) 1	2	3	4	5
(2) 2	1	(3) 7	4	(30) 6	5	3
4	(2) 7	5	(6) 1	(6) 2	3	6
(2) 5	(12) 3	4	7	1	(6) 6	2
3	(240) 5	2	6	4	1	(7) 7

241

(4) 2	(126) 7	3	6	(4) 4	(4) 1	5
6	(2) 4	(6) 7	1	(3) 5	2	(315) 3
(7) 4	2	(48) 6	(4) 5	1	3	7
3	(210) 6	2	4	(4) 7	5	(5) 1
7	5	(2) 1	2	3	(2) 6	4
(23) 1	(2) 3	5	(42) 7	6	4	(3) 2
5	1	4	3	2	7	6

242

(50) 2	5	(2) 1	3	(3) 4	(5040) 6	7
5	(1008) 6	4	(2) 2	1	(10) 7	3
(3) 3	7	6	4	2	1	5
1	(4) 3	7	(5) 5	(60) 6	2	4
(2) 6	4	(30) 3	(18) 1	(7) 7	5	2
(5880) 7	2	5	6	3	(4) 4	1
4	1	2	7	5	3	(6) 6

243

[756] 6	7	[28] 4	[5250] 3	5	2	1
3	6	7	[8] 1	[2] 2	4	5
[10] 2	4	[21] 3	6	1	5	7
4	[2] 2	6	5	7	[3] 1	3
[30] 7	5	[6] 1	2	3	[2] 6	4
5	[2] 1	2	[168] 7	4	[42] 3	[6] 6
1	3	5	4	6	7	2

244

[168] 7	4	[16] 3	5	2	1	[6] 6
6	[2] 2	5	[2] 3	[12] 4	[6] 7	1
[168] 2	7	[11] 6	1	3	[1] 4	5
1	3	2	[4] 4	5	[1] 6	[5] 7
3	[24] 1	4	6	[7] 7	5	2
4	[30] 5	[6] 1	[13] 7	6	[72] 2	3
[5] 5	6	7	[2] 2	1	3	4

245

[2] 1	[24] 4	3	2	[11] 5	6	[504] 7
2	[31] 5	7	1	6	3	4
[3] 7	3	[4] 2	[20] 5	4	[12] 1	6
4	2	6	[140] 7	[2] 3	5	[1] 1
[540] 6	7	5	4	1	2	[10] 3
5	6	1	[3] 3	[588] 7	4	2
3	[3] 1	4	6	2	7	5

246

[45] 3	5	[29] 6	2	7	4	1
[6] 7	3	4	[18] 1	[40] 2	[6] 6	5
1	[1] 7	3	6	4	5	[2] 2
[4] 2	6	[35] 5	7	1	[4] 3	4
6	[14] 2	1	[100] 4	5	7	[2] 3
[3] 4	1	7	5	[756] 3	2	6
[11] 5	4	2	3	6	1	7

247

[19] 2	7	[72] 6	4	[2] 5	3	[8] 1
4	6	3	[10] 1	2	5	7
[14] 3	[13] 4	2	[30] 5	[588] 7	[6] 1	6
5	[2] 2	1	6	4	[21] 7	3
6	1	5	7	3	[2] 2	4
[35] 7	[22] 3	4	2	[7] 1	6	[7] 5
1	5	[7] 7	3	6	4	2

248

[6] 6	[1] 1	2	[24] 4	[2] 5	3	[1120] 7
[15] 7	3	1	6	2	5	4
5	[21] 7	3	1	4	[15] 2	[14] 6
[2] 1	2	[72] 4	3	6	7	5
[120] 2	[1] 4	5	[15] 7	[56] 1	6	3
3	5	6	2	7	4	1
4	[13] 6	7	[5] 5	[3] 3	1	2

249

5: 5	288×: 1	6	6−: 2	8	4−: 3	7
3	2	6−: 7	8+: 5	14+: 6	8	9+: 1
8	840×: 7	1	3	7+: 2	5	6
8+: 1	3	5	8	13+: 7	6	2
7	1−: 6	5−: 8	126×: 1	3	10+: 2	3−: 5
3÷: 2	5	3	6	1	7	8
6	6−: 8	2	7	15×: 5	1	3

250

40×: 1	8	5	2−: 7	11+: 2	27+: 3	6
6−: 2	2÷: 1	1−: 7	5	3	6	8
8	2	6	6+: 1	5	7	3
3: 3	84×: 7	2	6	40×: 1	8	5
13+: 6	2−: 3	7−: 1	4÷: 2	8	70×: 5	7
7	5	8	11+: 3	84×: 6	2	1
90×: 5	6	3	8	7	1	2

251

42×: 2	1	3	7	14+: 6	8	5: 5
120×: 8	5	56×: 7	1	12×: 2	2÷: 3	6
3	16+: 7	8	19+: 5	1	6	3+: 2
2−: 5	3	6	8	14×: 7	2	1
7	15+: 8	2	6	2−: 5	1	18+: 3
36×: 1	6	5	1680×: 2	3	7	8
6	1−: 2	1	3	8	5	7

252

24×: 8	3	17+: 9	6+: 1	5	15+: 6	4
5−: 4	9	8	15+: 6	1	3	5
180×: 9	5	10368×: 3	8	6	4	6×: 1
3÷: 3	4	180×: 5	9	8	1	6
1	14+: 8	6	4	9	2−: 5	3
17+: 6	4÷: 1	4	60×: 5	3	576×: 8	9: 9
5	6	3÷: 1	3	4	9	8

253

8−: 1	15×: 5	3	14+: 6	8	36×: 4	9
9	1	26+: 5	3	192×: 6	8	4
3	6	9	4−: 1	8+: 4	17+: 5	19+: 8
4: 4	7−: 8	1	5	3	9	6
30+: 8	4	6	9: 9	1	3	5
1−: 5	216×: 3	4	8	54×: 9	6	1
6	9	8	1−: 4	5	3÷: 1	3

254

288×: 8	9	4	20×: 1	5	144×: 6	3
2÷: 3	15+: 5	9	4	1	8	32+: 6
6	1	3÷: 3	5: 5	5−: 4	9	8
24×: 4	6	1	17+: 8	9	360×: 3	5
1	4: 4	14+: 8	3−: 6	3	5	9
17+: 5	3	6	9	8	1: 1	4
9	2880×: 8	5	3	6	4	1: 1

255

3− 5	3456× 2	8	4	6	9	2− 7
8	70× 5	2	30× 6	20+ 4	7	9
2÷ 2	4	7	5	9	2− 8	6
210× 6	7	5	1134× 9	2÷ 8	4	2 2
3− 4	1− 8	9	7	2	3÷ 6	1− 5
7	9	14+ 6	8	2− 5	2	4
3− 9	6	2− 4	2	7	13+ 5	8

256

2÷ 2	84× 7	6	405× 9	5	29+ 8	2÷ 4
4	2	2− 5	7	9	6	8
23+ 9	3− 4	7	5	8	2	4− 6
6	486× 9	224× 8	4	7	5 5	2
8	6	9	26+ 2	4	7	315× 5
2− 7	17+ 5	4÷ 2	8	14+ 6	4	9
5	8	4	6	2	9	7

257

3− 7	4	3− 5	17+ 8	2268× 6	9	6− 2
1− 4	5	2	9	7	6	8
17+ 6	280× 8	7	5	5184× 9	2− 2	4
5	6	8	2	4	441× 7	9
4÷ 8	2	9	120× 6	5	4	7
126× 2	216× 9	4	224× 7	8	1− 5	6
9	7	6	4	2 2	3− 8	5

258

3780× 4	9	3	1680× 6	8	5	7
7	4 4	18+ 6	2− 5	17+ 9	8	6− 3
5	8	4	3	1− 6	7	9
15+ 3	5	7	2÷ 8	4	27+ 9	6
15+ 6	4− 3	3− 5	63× 9	7	4	8
9	7	8	60× 4	3	11+ 6	5
432× 8	6	9	7 7	5	7+ 3	4

259

2÷ 6	3	2− 7	5	288× 9	8	4
14+ 4	45× 5	9	5− 3	8	22+ 6	7
3	4 4	18+ 8	13+ 7	6	9	270× 5
7	17+ 9	4	224× 8	5 5	3	6
17+ 9	8	6	4	7	180× 5	3
8	2− 7	5	2÷ 6	3	4	9
11+ 5	6	3÷ 3	9	19+ 4	7	8

260

4032× 3	6	40× 8	5	16+ 9	7	2− 4
7	3 3	17+ 9	8	1− 5	4	6
4	8	27+ 5	6	7	3÷ 9	3
14+ 6	5	4	16+ 9	72× 8	3	7 7
8	20+ 4	6 6	7	3	35+ 5	9
45× 5	9	7	13+ 3	4	6	8
9	4− 7	3	2− 4	6	8	5

261

240× 6	5	2− 3	1	13+ 4	9	8 8
8	12× 1	4	2− 3	5	54× 6	9
3÷ 9	3	120× 6	4	3− 8	5	1
3	2− 6	5	9 9	24× 1	8	1− 4
4− 1	8	8− 9	14+ 6	3	108× 4	5
5	1152× 4	1	8	9	3	18× 6
4	9	8	1− 5	6	1	3

262

3÷ 3	4− 1	5	7− 8	12× 4	19+ 6	9
9	48× 6	8	1	3	4	1− 5
3− 6	9	14+ 1	162× 3	19+ 8	5 5	4
1− 4	3	9	6	5	7− 8	1 1
3− 5	8	4	9	6	1	2÷ 3
9+ 8	1− 4	3	20× 5	8− 1	216× 9	6
1	11+ 5	6	4	9	3	8

263

17+ 4	144× 8	3	3÷ 1	26+ 9	6	6+ 5
9	4	6	3	5	24× 8	1
9+ 5	36× 9	1	4	6	3	23+ 8
1	12× 3	4	3− 5	8	9 9	6
3	48× 1	8	6	12× 4	6+ 5	9
20+ 8	6	225× 5	17+ 9	3	1	1− 4
6	5	9	8	4÷ 1	4	3

264

140× 4	1	7	40× 8	5	9+ 2	6
5− 7	5	96× 2	6	14+ 8	11+ 4	1
2	4	1	120× 5	6	7	21+ 8
9+ 1	2	6	4	6− 7	8	5
8	24× 6	4	70× 2	1	2− 5	7
240× 5	1− 7	8	1	2÷ 2	24× 6	4
6	8	5	7	4	1	2 2

265

42× 7	2− 4	6	3− 2	40× 5	7− 1	8
6	16+ 1	2	5	8	28× 7	4
1− 1	5	8	840× 6	6− 7	25+ 4	2
2	7	5	4	1	8	6 6
160× 8	13+ 2	7	4÷ 1	4	6	5
5	48× 8	4	1− 7	4− 6	2÷ 2	1
4	6	1	8	2	2− 5	7

266

112× 7	9+ 5	180× 6	2÷ 8	4	1− 3	2
8	4	2	3	5	30× 6	10+ 7
2	6720× 6	4	336× 7	8	5	3
1− 4	3	5	6	14× 7	2	8 8
14+ 3	7	8	4− 2	6	11+ 4	960× 5
6	336× 8	3	3− 5	2	7	4
5	2	7	1− 4	3	8	6

267

27× 1	3	2− 6	4	6− 8	2	54× 9
9	14+ 6	8	48× 1	12× 4	3	2
48× 6	4	5− 9	8	1− 2	1	3
2	4÷ 8	4	6	3	19+ 9	5− 1
2÷ 8	2	3÷ 1	3	8− 9	4	6
4	9 9	6× 3	17+ 2	1	6	20+ 8
3 3	1	2	9	6	8	4

268

72× 4	6	1	3	2− 8	20+ 2	9
3÷ 9	24× 4	3	16× 2	6	8	1
3	6− 9	2	1	48× 4	6	96× 8
1 1	3	5− 4	8	2	13+ 9	6
16+ 6	8	9	10+ 4	1	3	2
2	15+ 1	8	6	21+ 9	4÷ 4	12× 3
4÷ 8	2	6	9	3	1	4

269

35+ 5	8	7	11+ 3	11+ 6	1	4	6+ 2
10+ 3	7	8	6	5+ 2	11+ 5	1	4
7− 8	1	6	2	3	13+ 4	5	15+ 7
5+ 4	10+ 2	1	12+ 5	7	3	13+ 6	8
1	3	6+ 2	4	7− 8	6	7	1− 5
15+ 7	5	15+ 4	7− 8	1	1− 2	3	6
2	6	5	1	17+ 4	15+ 7	8	2− 3
6	14+ 4	3	7	5	8	2 2	1

270

15+ 8	6	9+ 4	3	27+ 5	2	1	7
4 4	1	2	9+ 5	3	1− 6	7	8
1− 5	14+ 7	3	4	3+ 2	1− 8	14+ 6	1
6	4	7+ 5	2	1	7	8	3 3
20+ 7	5	7− 8	1	3− 6	3	2− 4	3− 2
6+ 3	8	13+ 7	6	18+ 4	1	2	5
1	2	5− 6	7 7	8	5	7+ 3	4
5+ 2	3	1	15+ 8	7	15+ 4	5	6

271

7 7	28+ 8	1	8+ 5	3	17+ 4	14+ 2	6
5+ 2	4 4	3	7	5	8	6	7− 1
1	2	2− 4	3	6	5 5	21+ 7	8
8+ 3	1	2	17+ 4	7	6	8	3− 5
4	16+ 3	12+ 7	6	7− 8	1	10+ 5	2
6	7	5	7− 8	5+ 2	3	1	4
19+ 5	6	8 8	1	13+ 4	2	17+ 3	7
8	1− 5	6	3+ 2	1	7	4	3

272

27+ 7	2	1 1	4	8+ 3	5	33+ 6	8
3− 1	3	6	5	4	7	8	8+ 2
4	7− 8	3+ 2	1	10+ 7	3	1− 5	6
11+ 5	1	5+ 3	2	2− 6	8	4	4− 7
6	12+ 7	5	7− 8	1	6+ 4	2	3
3− 8	5	1− 4	3	4− 2	6	6− 7	1
9+ 2	4	13+ 7	6	7− 8	1	12+ 3	5
3	2− 6	8	2− 7	5	3+ 2	1	4

273

8+ 7	13+ 3	1– 6	5	3+ 2	1	22+ 8	12+ 4
1	6	4	1– 2	3	7	5	8
14+ 4	5– 7	7– 1	8	10+ 6	3	2	32+ 5
5	2	7+ 3	4	1	8	7 7	6
3	15+ 8	7	3– 1	4	5	6	2
2	7– 1	8	21+ 6	5	14+ 4	3	7
13+ 8	5	3– 2	3	7	4– 6	5+ 4	1
10+ 6	4	5	1– 7	8	2	4+ 1	3

274

4+ 1	3	3+ 2	15+ 8	7	1– 6	5	19+ 4
30+ 8	5	1	7+ 3	4	7	6	2
6	6+ 1	3	2	1– 5	4	15+ 7	8
2	4	5	6 6	7– 8	4– 1	4– 3	7
23+ 3	2	18+ 4	7	1	5	14+ 8	6
7	8 8	6	1	1– 3	2	9+ 4	5
5	6	1– 7	6+ 4	2	7– 8	1	4+ 3
3– 4	7	8	1– 5	6	1– 3	2	1

275

6+ 4	2	12+ 5	7	5– 1	6	23+ 3	8
31+ 2	6	7+ 4	3	7– 8	1	7	5
7	3 3	7– 1	8	1– 2	15+ 4	5	6
8	1	2	5	3	15+ 7	10+ 6	5+ 4
26+ 5	7	18+ 3	12+ 2	6	8	4	2
3	8	6	1	4	14+ 5	2	7
6	5	12+ 8	4	4– 7	3	7– 1	5+ 2
3– 1	4	13+ 7	6	7+ 5	2	8	3

276

13+ 6	15+ 7	12+ 3	6+ 2	4	7– 1	8	15+ 5
7	8	6	3	8+ 1	16+ 5	4	2
6+ 1	5	3+ 2	4	3	6 6	7	8
18+ 8	3	1	3– 5	2	2– 4	6	11+ 7
2	1 1	3– 7	7– 8	41+ 6	4– 3	14+ 5	4
5	16+ 2	4	1	8	7	3	6
7+ 4	6	8	7 7	5	10+ 2	4+ 1	3
3	4	5	6	7	8	3+ 2	1

277

15+ 7	11+ 5	6	8 8	3+ 2	1	1– 4	3
2	27+ 7	4	3	5	8	1– 6	20+ 1
6	3+ 1	2	14+ 7	7+ 4	3	5	8
3+ 1	2	4– 3	5	7– 8	2– 6	7	4
18+ 3	6	7	2	1	4	15+ 8	5
4	5– 3	21+ 8	6	7	4– 5	1	2
5	8	1 1	16+ 4	3	7	2	1– 6
4– 8	4	12+ 5	1	6	5+ 2	3	7

278

17+ 5	8	9+ 2	3– 4	4– 3	7	1– 6	4+ 1
7– 8	4	1	7	8+ 2	9+ 6	5	3
1	20+ 5	4	2	6	3	15+ 7	8 8
17+ 6	1	3	5	4	2	8	15+ 7
4	5+ 3	21+ 8	1	7	5	2	6
7	2	1– 5	14+ 6	1 1	7– 8	13+ 3	6+ 4
5+ 3	13+ 7	6	8	13+ 5	1	4	2
2	6	4– 7	3	8	4 4	1	5

279

2– 2	4+ 1	2– 4	21+ 8	7	5	9+ 6	3
4	3	6	5+ 2	1	17+ 8	7	2– 5
11+ 5	7– 8	1	3	17+ 6	4	2	7
6	6+ 2	3	1	13+ 8	7	15+ 5	4
7– 1	21+ 6	8	7	5	9+ 3	4	2
8	2– 5	7	2– 6	4	2	10+ 3	7– 1
10+ 3	11+ 7	12+ 5	4	6+ 2	6	1	8
7	4	2	5	3	1	14+ 8	6

280

13+ 6	1	22+ 7	2	8	5	7+ 4	3
6+ 2	6	1– 5	7– 8	1	8+ 3	21+ 7	11+ 4
1	5 5	6	16+ 4	3	2	8	7
3	6+ 4	2	5	6– 7	1	6	6– 8
16+ 5	7+ 3	4	7	2– 6	8	9+ 1	2
7	25+ 8	9+ 1	3 3	6+ 2	4	5	5– 6
4	2	8	11+ 6	5	7 7	3	1
8	7	14+ 3	1	4	6	3– 2	5

281

21+ 7	6	8+ 3	2	9+ 5	4	7– 8	1
5	1	2	18+ 8	4	21+ 7	9+ 6	3
1	2	6+ 5	10+ 3	6	8	4 4	15+ 7
15+ 3	4	1	5	2	6	12+ 7	8
2	15+ 3	4	21+ 7	7– 8	1	5	10+ 6
6	18+ 5	8	1	7	2	2– 3	4
20+ 8	7	6	4	17+ 3	2– 5	1	9+ 2
4	8	7	6	1	3	2	5

282

12+ 5	6+ 2	3	1	19+ 7	4	23+ 8	6
7	9+ 4	8+ 5	5+ 3	2	17+ 6	1	8
8+ 4	5	1	2	6	8	3	2– 7
1	3	2	14+ 4	21+ 8	7	6	5
6– 2	7– 1	8	6	4	12+ 5	7	9+ 3
8	17+ 6	4	20+ 7	9+ 3	1	5	2
21+ 6	8	7	5	1	1– 3	2	4
3 3	7	14+ 6	8	5	2	3– 4	1

283

3– 2	7– 1	8	6+ 4	7+ 3	11+ 6	5	13+ 7
5	8 8	3– 1	2	4	17+ 3	7	6
9+ 6	2	4	4– 1	5	7	19+ 8	3
1	12+ 6	3	15+ 5	21+ 7	8+ 2	4	8
24+ 4	3	6 6	7	8	3– 1	2	11+ 5
8	7	13+ 5	3	6	4	1	2
14+ 7	5	2	6	7– 1	8	3	11+ 4
3	4	15+ 7	8	7+ 2	5	6	1

284

14+ 1	5	4	21+ 8	7	6	18+ 2	3
15+ 5	4	6+ 1	2	3	7	6	7– 8
6	15+ 7	12+ 3	4	9+ 2	5	7– 8	1
4	8	5	3– 3	6	2	1	18+ 7
1– 3	2	8+ 6	21+ 5	5+ 1	7– 8	7	4
2– 8	6	2	7	4	1	8+ 3	5
9+ 2	2– 3	21+ 7	1	8	4 4	9+ 5	4– 6
7	1	8	6	8+ 5	3	4	2

285

16+ 3	8	1– 2	1	9+ 5	4	1– 7	6
5	1– 2	10+ 7	3	7+ 1	6	10+ 4	7– 8
21+ 8	3	17+ 4	5	16+ 7	2	6	1
6	1– 4	3	8	8+ 2	7	6+ 1	16+ 5
7	7+ 1	7+ 5	2	6	12+ 8	3	4
1	5	18+ 8	6	4	3	2	7
2– 2	19+ 7	6	4– 4	8	1	8+ 5	3
4	6	6– 1	7	16+ 3	5	8	2 2

286

17+ 6	7	4– 1	14+ 8	6+ 3	11+ 4	2	5
4	18+ 6	5	2	1	19+ 8	18+ 7	3
7– 1	3	7	4	2	5	6	8
8	8+ 1	2	7+ 3	4	13+ 6	12+ 5	7
7+ 2	4	3	1– 6	5	7	7– 8	1
3	2	13+ 8	5	21+ 7	6+ 1	17+ 4	6
12+ 7	5	5– 6	1	8	2	3	4
3– 5	8	3– 4	7	6	3	3+ 1	2

287

1– 4	2– 1	8+ 7	7+ 3	6– 8	5– 2	1– 6	5
5	3	1	4	2	7	15+ 8	14+ 6
7+ 6	15+ 2	3	1	5	4	7	8
1	9+ 4	3– 2	5	29+ 6	8	4– 3	7
6– 2	5	14+ 6	8	7	3	5+ 4	1
8	18+ 6	4	7	1	5	3+ 2	1– 3
4– 7	15+ 8	13+ 5	4– 2	7+ 3	5– 6	1	4
3	7	8	6	4	1	3– 5	2

288

3– 2	1– 3	4	14+ 8	6	6– 7	1	2– 5
5	6– 1	13+ 2	16+ 4	7	15+ 6	8	3
11+ 4	7	8	3	5	1	5– 2	21+ 6
6	15+ 2	1	5	4	3	7	8
1	1– 6	5	9+ 2	5– 3	4– 8	4	7
18+ 3	2– 4	6	7	8	3– 2	5	11+ 1
7	8	4+ 3	1	7+ 2	5	6	4
13+ 8	5	14+ 7	6	1	4 4	5+ 3	2

289

840× 7	8	5	24× 3	4	24× 6	2352× 1	2× 2
90× 6	5	3	2	280× 8	4	7	1
3	192× 2	8	1	5	7	6	20× 4
16× 2	3	4	42× 6	7	1	8	5
8	1	168× 6	7	1200× 2	5	4	18× 3
28× 1	4	7	288× 8	3	2	5	6
4	7	12× 1	20× 5	6	24× 3	2	56× 8
5 5	6	2	4	1	8	3	7

290

2880× 4	5	105× 7	3	18× 6	1	96× 8	2
210× 2	4	6	5	3	256× 8	336× 7	1
5	7	3	2 2	8	4	1	6
3	6× 1	2	20× 4	5	42× 7	6	8
1	6	320× 5	8	2	3	28× 4	7
12× 6	2	8	1	35× 7	5	60× 3	4
56× 7	24× 8	672× 4	6	1	12× 2	5	30× 3
8	3	1	7	4	6	2	5

291

40× 8	4200× 5	7	3	2 2	96× 6	1	4
5	24× 3	8	2	28× 1	4	6 6	126× 7
6× 2	64× 8	1	5	4	7	3	6
3	2	30× 5	4	144× 6	8	280× 7	1
168× 7	4	6	1	6720× 8	3	5	6× 2
4	1	12× 2	6	7	10× 5	8	3
6	126× 7	224× 4	8	3	1	2	40× 5
1	6	3	7	5	2	4	8

292

43200× 8	5	14× 2	7	1	1152× 6	3	4
7× 7	6	5	1	8064× 3	8	4 4	2
1	12× 2	3	600× 4	5	7	6	8
2	1	4	3	6	5	8	84× 7
3	576× 4	240× 8	5	2352× 7	1	2	6
7840× 4	3	1	6	8	2	7	5 5
5	7	6	8	16× 2	60× 4	1	3
6 6	8	7	2	4	3	5	1

293

13440× 5	7	8	1	120× 6	24× 4	2	3
3	96× 8	84× 7	2	5	1	4	6 6
2	3	4	6	168× 7	40× 8	5	14× 1
8	42× 1	3	40× 5	4	6	7	2
7	2	5 5	8	1	6× 3	168× 6	4
120× 6	5	1	32× 4	8	2	24× 3	7
1	4	288× 6	6× 3	2	35× 7	8	40× 5
4	6	2	21× 7	3	5	1	8

294

3 3	768× 2	8	30× 6	5	24× 1	420× 4	7
4	6	336× 7	2	8	3	1	5
2	8	1	24× 4	6	56× 7	2880× 5	3
30× 5	3	10× 2	84× 7	4	8	6	1
6	160× 4	5	1	3	392× 2	7	8
1	5	24× 3	8	7	4	6× 2	6
8	33600× 7	4	5	1	6 6	3	2
42× 7	1	6	3	2	5	8	4 4

295

5376× 2	1	8	280× 5	7	150× 6	12× 3	4
6	14× 7	12× 3	4	8	5	1	6× 2
8	2	288× 6	1	112× 4	7	5	3
7	3	2	8	60× 5	4	336× 6	1
2400× 3	20× 5	4	6	2	6× 1	8	7
5	24× 6	1	147× 7	3	2	192× 4	40× 8
1	4	7	3	12× 6	8	2	5
4	8	5	2	1	3	42× 7	6

296

7 8	2 2	15 7	24 6	30 5	4 4	1	42 3
1	60 3	8	4	6	1680 5	21 7	2
4	5	10 1	2	8	6	3	7
25 6	3 1	2	3	4	7	3 5	8
5	8	6	448 7	2	7 3	480 4	1
84 3	7	4	8	1	2	48 6	5
9 2	4	90 3	6 5	4 7	1	8	6
7	6	5	1	3	6 8	2	4

297

(126) 7	(6) 6	1	(2) 5	3	(2) 4	2	(7) 8
6	(768) 8	2	(1) 3	4	(1960) 5	7	1
3	1	4	(3) 2	5	7	(14) 8	6
(3) 1	2	3	4	(15) 7	8	(1) 6	5
(11) 4	5	(2016) 6	1	8	(3) 2	(420) 3	7
2	(3) 3	8	7	(2) 1	6	5	4
(56) 8	7	(140) 5	6	2	(5) 1	(24) 4	3
(5) 5	4	7	(48) 8	6	3	1	2

298

(90) 6	5	(42) 3	1	7	2	(2) 8	4
3	(2) 2	(192) 6	4	(15) 5	(7) 8	1	(11) 7
(15) 8	7	2	6	4	(4) 5	3	1
(5) 2	3	4	(15) 8	(3) 6	1	(35) 7	(2) 5
(2) 4	8	(35) 1	7	2	(42) 6	5	3
(6) 1	4	5	(144) 2	3	7	(288) 6	8
(210) 5	1	7	3	8	(2) 4	2	6
7	6	(17) 8	5	1	3	(2) 4	2

299

(60) 3	(10) 5	(14) 6	2	(3) 1	(15) 8	7	(12) 4
5	4	1	6	3	(1050) 7	(24) 8	2
4	(112) 7	2	1	8	5	3	6
(15) 7	8	(1) 4	(5) 3	5	6	(2) 2	1
(11) 2	3	5	8	(32) 4	1	(13) 6	7
(7) 1	6	(280) 8	5	7	2	(4) 4	(120) 3
8	(6) 2	3	(196) 7	(3) 6	4	1	5
(6) 6	1	7	4	2	(2) 3	5	8

300

(2) 6	3	(2) 1	(15) 7	8	(26) 2	5	4
(30) 5	1	2	(12) 4	(2304) 6	(126) 3	(8) 8	7
1	2	3	5	4	6	7	8
3	(224) 7	4	1	2	8	6	(1) 5
2	4	(336) 7	(7) 8	1	(15) 5	3	6
(33) 4	(150) 5	8	6	(13) 3	7	(7) 2	1
8	6	5	3	7	(16) 1	4	2
7	8	6	(2) 2	5	4	1	3

301

(63) 7	3	(100) 4	5	(5) 1	6	(64) 8	(2) 2
3	(3) 2	5	(1) 7	6	8	1	4
(9) 8	1	(144) 3	6	(13) 4	(5) 7	(60) 2	5
1	(15) 7	8	4	5	2	6	(2) 3
(20) 4	8	(3) 1	2	(12) 7	3	(1575) 5	6
5	(48) 6	(15) 7	8	2	(4) 4	3	(21) 1
2	4	(9) 6	1	3	5	7	8
(30) 6	5	2	(24) 3	8	1	4	7

302

(2) 4	(3) 2	1	(21) 7	(72) 6	3	(12) 5	(40) 8
2	(6) 6	3	1	4	(96) 8	7	5
(42) 7	(28) 8	4	3	(7) 5	6	2	(3) 1
6	(1) 4	5	8	2	(6) 7	1	3
(7) 8	5	(12) 2	(24) 4	(15) 7	1	(36) 3	6
1	3	7	6	8	(1) 5	4	2
(2) 5	(42) 1	(15) 8	2	(6) 3	(192) 4	6	(11) 7
3	7	6	5	1	2	8	4

303

(896) 8	7	(3) 2	6	(9) 5	4	(9) 3	1
2	8	(7) 7	(84) 3	(180) 6	5	1	4
(3) 1	(24) 2	3	4	7	6	(30) 5	(3) 8
3	1	4	(10) 2	(15) 8	7	6	5
(12) 4	3	(12) 6	5	1	(15) 8	7	(16) 2
(16) 7	4	5	1	(6) 3	2	8	6
(17) 6	5	(7) 8	(56) 7	(2) 4	1	(6) 2	3
5	6	1	8	2	(84) 3	4	7

304

(270) 3	6	(20) 5	1	4	(12) 2	(3) 7	(15) 8
5	3	(144) 2	6	(15) 8	1	4	7
(3) 2	1	3	4	7	6	(200) 8	5
(7) 1	(5) 2	(4) 4	(35) 7	(14) 6	8	5	(2) 3
8	7	1	5	(45) 3	(11) 4	(2) 2	6
(4) 4	(20) 8	(16) 6	3	5	7	1	(8) 2
7	5	8	2	(6) 1	3	(90) 6	4
(10) 6	4	(1) 7	8	2	5	3	1

305

(252) 7	6	3	2	(16) 5	4	(7) 8	1
(8) 5	(3) 1	2	(2) 8	4	7	(126) 6	3
1	2	(14) 5	(105) 3	(14) 6	8	(3) 4	7
(5) 3	4	1	5	(112) 2	(6) 6	7	(48) 8
2	(2) 3	4	7	8	(3) 1	(5) 5	6
(336) 6	5	(7) 8	1	7	2	(14) 3	4
8	7	(252) 6	(12) 4	3	(4) 5	1	2
(2) 4	8	7	6	(6) 1	3	2	5

306

(144) 3	(3) 8	5	(3) 1	(14) 7	(4) 4	(19) 2	6
8	(1120) 4	(7) 7	3	2	1	6	5
6	7	(6) 3	2	1	(160) 5	4	8
2	1	(2) 8	4	(240) 6	(12) 3	(35) 5	7
4	(432) 3	6	(5) 5	8	2	(8) 7	1
1	(3) 2	4	6	5	7	(8) 8	(1) 3
5	6	(14) 2	7	(30) 3	8	1	4
(7) 7	5	1	8	4	(36) 6	3	2

307

(2) 2	(3) 6	(7) 8	1	(3) 7	4	(40) 5	(2) 3
1	3	(3) 4	7	(4) 2	6	8	5
(12) 5	7	(2) 2	4	(24) 3	8	(6) 6	(10752) 1
(2) 8	(60) 2	6	5	(4) 4	(3) 1	3	7
4	(21) 5	3	(21) 8	6	7	(8) 1	2
(126) 6	8	5	(6) 2	1	3	7	4
7	(4) 4	1	(5) 3	8	(7) 5	2	6
3	1	(42) 7	6	(3) 5	2	4	8

308

(10080) 3	(28) 7	(120) 5	4	(7) 2	(7) 1	8	(6) 6
2	4	6	(2) 1	5	(14) 7	3	(40) 8
6	(4) 1	2	3	(224) 7	8	4	5
1	2	(3) 3	(240) 5	8	4	(13) 6	7
7	5	(2) 4	8	6	(1) 3	2	(10) 1
(60) 4	8	(6) 1	7	(12) 3	6	5	2
5	3	(29) 8	6	1	2	(28) 7	4
(336) 8	6	7	2	4	5	1	3

309

(12) 4	(2) 3	5	(4) 1	(5) 6	(1) 7	8	(6) 2
3	(8) 2	(4) 7	4	1	(40) 5	(3) 6	8
1	4	3	(5) 5	(20) 7	8	2	(2520) 6
(15) 6	(3) 5	2	(4) 7	8	(2) 3	4	1
2	(1) 7	8	3	5	6	1	(20) 4
7	(15) 8	(7) 1	6	(2) 2	4	3	5
(40) 8	6	(24) 4	(6) 2	3	1	5	(10) 7
5	1	6	(2) 8	4	(2) 2	7	3

310

(3) 6	2	(21) 4	1	8	(336) 3	7	(5) 5
1	4	3	(42) 6	7	(175) 5	2	8
(54) 3	6	(8) 8	(2) 4	2	7	5	(5376) 1
(28) 7	3	(4) 2	8	(5) 5	(7) 1	(5) 6	4
4	(105) 5	7	3	(2) 6	8	1	2
(4) 2	(30) 1	6	(10) 5	3	(4) 4	8	7
8	(7) 7	5	2	(9) 1	6	4	(2) 3
(5) 5	(7) 8	1	(28) 7	4	2	3	6

311

(75) 3	5	(1) 2	1	(11) 7	4	(1536) 8	6
5	(84) 7	(2) 3	(3) 2	(3) 1	(48) 6	4	8
(8) 1	4	6	5	3	8	(1) 7	(2) 2
7	3	(140) 4	(21) 8	(16800) 5	(6) 2	6	1
(2) 4	8	5	7	6	1	(5) 2	3
(7) 8	1	7	6	2	3	(1) 5	4
(24) 6	2	1	(11) 4	8	5	(4) 3	7
2	(2) 6	8	3	4	7	1	(5) 5

312

(2) 2	(2) 3	6	(336) 1	8	(1800) 4	(280) 7	5
4	7	3	2	(22) 1	5	6	8
(36) 6	(1) 1	(3) 7	4	2	8	5	3
1	6	(4) 4	(1050) 5	7	3	8	(12) 2
(21) 7	(4) 4	8	(3) 3	5	(5) 6	2	1
3	(8) 2	5	(15) 8	6	1	4	(1) 7
(320) 8	5	1	7	(14) 4	(2) 2	3	6
(5) 5	8	(3) 2	6	3	7	(3) 1	4

313

(9) 3	(448) 1	4	2	8	(1) 7	6	(5) 5
2	(3) 3	6	(10) 1	7	(140) 5	4	(7) 8
4	(210) 5	2	6	3	(48) 8	7	1
(280) 8	(2) 4	3	7	(3) 2	1	(1680) 5	6
5	8	(4) 1	4	6	2	3	7
1	7	(2) 5	3	(13) 4	6	(7) 8	2
(252) 6	(2) 2	(1) 7	(3) 8	5	3	1	4
7	6	8	(6) 5	1	(4) 4	(1) 2	3

314

(16) 6	(28) 4	7	1	(21) 8	(3) 3	(7) 2	5
7	3	2	5	6	1	(4) 8	4
(1) 5	(3) 2	6	(42) 3	(11) 4	7	(7) 1	8
4	(56) 8	(8) 3	7	2	(11) 5	(7) 6	1
(3) 1	7	5	(48) 8	3	6	(2) 4	2
3	(7) 1	8	2	(9) 5	4	(13) 7	6
(96) 8	6	(28) 1	4	7	(3) 2	5	(3) 3
2	(120) 5	4	6	1	(8) 8	(4) 3	7

315

(18) 7	4	(2) 1	2	(48) 6	(120) 5	(24) 3	8
(15) 5	7	(3) 2	1	8	6	4	(4) 3
4	(1) 3	5	(6) 8	2	(10) 1	(6) 6	7
3	2	(2) 4	(210) 6	7	8	1	(3) 5
1	(5) 6	8	5	(4) 4	(1176) 3	7	2
2	1	(2) 3	(9) 4	5	7	8	(7) 6
(3) 8	5	6	(147) 7	(2) 3	(8) 4	2	1
(2) 6	8	7	3	1	2	(9) 5	4

316

(11+) 9	2	(16+) 7	8	1	(2−) 5	3	(2−) 4	6
(2−) 2	(7+) 3	(1−) 6	7	(11+) 5	(8−) 9	1	(17+) 8	(4−) 4
4	1	3	(16+) 2	6	(17+) 7	(5) 5	9	8
(4−) 1	(14+) 4	2	3	7	6	(17+) 8	(5) 5	(23+) 9
5	(19+) 6	8	4	3	1	9	(11+) 2	7
(4−) 3	8	(8−) 9	(7+) 1	4	2	(3−) 7	6	5
7	5	1	(16+) 6	(17+) 9	8	4	3	2
(14+) 8	(5−) 9	4	5	2	3	(1−) 6	7	(4+) 1
6	(2−) 7	5	(9) 9	(4−) 8	4	(3+) 2	1	3

317

(18+) 9	7	(23+) 3	(4−) 4	8	(11+) 5	6	(3+) 1	2
2	(5) 5	4	(4+) 3	(15+) 6	9	(16+) 7	8	1
8	6	2	1	(8−) 9	(12+) 4	3	5	(21+) 7
(3−) 4	(3−) 3	6	(11+) 8	1	(2) 2	(12+) 5	7	9
7	(6−) 8	1	2	(16+) 3	6	(9) 9	(2−) 4	5
(8+) 3	2	(24+) 9	(5) 5	7	(8+) 1	4	6	(14+) 8
5	(8−) 9	8	(16+) 7	(4) 4	3	(5+) 1	2	6
(11+) 6	1	7	9	(3−) 5	8	2	(3) 3	(16+) 4
1	4	(13+) 5	6	2	(1−) 7	8	9	3

318

(8+) 1	(12+) 3	5	4	(2−) 7	(17+) 9	2	6	(17+) 8
4	(1−) 6	7	(17+) 2	5	(8+) 1	(17+) 9	8	3
3	(3−) 7	4	9	(6+) 1	2	(3−) 8	5	6
(4−) 8	4	(8+) 2	6	3	5	(2−) 1	(16+) 9	7
(16+) 7	(8−) 9	6	(18+) 8	2	(4−) 4	3	(8+) 1	(9+) 5
9	1	3	7	(14+) 6	8	5	2	4
(13+) 6	5	(8−) 9	1	8	(3−) 7	4	(9+) 3	2
(3−) 5	2	(20+) 8	3	9	(13+) 6	7	4	(8−) 1
2	(7−) 8	1	(12+) 5	4	3	(13+) 6	7	9

319

(12+) 2	6	(16+) 9	(5−) 8	(14+) 5	(8−) 1	(1−) 7	(12+) 4	(3) 3
(4−) 5	4	7	3	2	9	6	8	(5−) 1
9	(23+) 5	(6+) 2	4	7	(11+) 3	8	(18+) 1	6
3	7	8	(8−) 1	9	5	4	6	2
(28+) 6	8	3	(11+) 5	4	2	(8−) 1	9	(24+) 7
1	(3) 3	6	(19+) 7	8	4	(3−) 2	5	9
4	(6+) 1	5	(22+) 9	6	7	(1−) 3	2	8
(1−) 8	(8−) 9	1	(8+) 2	(4+) 3	(2−) 6	(12+) 5	7	(9+) 4
7	(6+) 2	4	6	1	8	(12+) 9	3	5

320

(30+) 6	9	7	8	(3+) 2	1	(12+) 4	5	3
(9+) 4	(8+) 6	(3+) 1	2	(16+) 3	8	5	(5−) 9	(8+) 7
5	2	(4−) 3	7	(15+) 6	(17+) 9	8	4	1
(1−) 7	8	(10+) 2	3	9	(9+) 5	1	(2−) 6	4
(2−) 2	4	5	(25+) 6	8	7	3	(8−) 1	9
(17+) 8	(17+) 3	6	(8−) 9	1	4	(16+) 2	7	(1−) 5
9	(9+) 1	8	(11+) 4	5	2	7	(11+) 3	6
3	5	(5+) 4	1	(3−) 7	(15+) 6	9	8	(2) 2
(8+) 1	7	(4−) 9	5	4	(3−) 3	6	(10+) 2	8

321

8– 9	1– 6	7	30+ 3	8	5	3– 1	4	7+ 2
1	17+ 8	7+ 4	3– 7	9	3– 3	14+ 6	5– 2	5
6– 2	9	3	4	5	6	8	7	2– 1
8	5– 7	2	20+ 9	4	1	11+ 5	6	3
4+ 3	1	3– 5	2	6	4– 8	4	16+ 9	7
3– 6	3	8– 9	1	19+ 7	6+ 4	2	13+ 5	8
4 4	3+ 2	1	20+ 5	3	9	21+ 7	8	6
1– 5	4	22+ 8	6	2	7	6– 3	8– 1	9
2– 7	5	6	8	3+ 1	2	9	7+ 3	4

322

7 7	22+ 8	2	9+ 1	3	5	19+ 6	9	4
17+ 4	7 7	1	2	9+ 5	17+ 8	9	9+ 6	3
1	3	7 7	9	4	4– 6	2	3– 8	5
3– 6	5	4	20+ 7	8	8– 9	1	10+ 3	2
3	15+ 6	9	5	6– 7	6+ 2	15+ 8	4	1
8+ 2	4	11+ 6	17+ 8	1	3	7	26+ 5	9 9
22+ 9	2	5	3	6	1	4	7	28+ 8
5	8– 9	15+ 8	4	21+ 2	7	3	1 1	6
8	1	3	6	9	4	5	2	7

323

18+ 9	2	7	14+ 8	1	12+ 5	3	4	18+ 6
6+ 2	4+ 3	1– 6	7	5	8– 9	1	17+ 8	4
4	1	7+ 3	2	19+ 6	7	33+ 5	9	8
3– 1	4	2	14+ 3	7 7	6	8	4– 5	9
16+ 5	14+ 6	8	4	2– 3	1	9	9+ 2	7
3	8	8– 9	1	2– 4	2	7	11+ 6	5
1– 7	4– 5	1	6	17+ 9	8	4	10+ 3	6+ 2
8	9	9+ 4	5	19+ 2	13+ 3	6	7	1
18+ 6	7	5	9	8	4	3+ 2	1	3

324

9+ 5	4	8– 1	11+ 8	3	11+ 9	2	16+ 7	6
3+ 1	24+ 8	9	4– 6	2	5 5	16+ 7	4	3
2	9	5– 8	3	4 4	20+ 7	6	5	3– 1
15+ 9	7	5+ 3	2	16+ 5	8	1	6	4
6	16+ 5	7	4	8– 1	3	8 8	17+ 9	7+ 2
13+ 3	2	18+ 6	7	9	5+ 1	5– 4	8	5
8	8+ 1	2	5	1– 6	4	9	10+ 3	7
14+ 4	6 6	5	8– 9	7	6+ 2	3	1	1– 8
7	3	4 4	1	2– 8	6	7+ 5	2	9

325

3– 8	10+ 7	3	16+ 6	5+ 4	1	8– 9	7+ 2	5
5	21+ 6	1– 9	2	8	3 3	1	8+ 7	3– 4
4 4	2	8	6+ 5	4– 3	3– 9	6	1	7
4– 3	4	6 6	1	7	3– 5	8	14+ 9	4– 2
7	9	12+ 4	8	1 1	5+ 2	3	5	6
8+ 6	5– 8	6+ 1	3	2	1– 7	1– 5	12+ 4	10+ 9
2	3	13+ 7	14+ 9	5	6	4	8	1
8– 9	1	5	1– 7	6	2– 4	2	9+ 3	11+ 8
1	11+ 5	2	4	24+ 9	8	7	6	3

326

17+ 8	12+ 6	6+ 2	1	3	15+ 7	1– 9	1– 5	4
6	4	6– 1	2 2	5	3	8	1– 9	24+ 7
3	2	7	2– 6	4	6+ 5	1	8	9
2– 5	7	18+ 9	15+ 4	2	6	3	22+ 1	8
1– 4	3	6	3– 5	8	7– 9	2	7	1 1
4– 1	5	3	30+ 7	9	8	6	4	2
18+ 2	8– 1	12+ 8	23+ 9	6	11+ 4	7	3– 3	5
7	9	4	8	4+ 1	2	1– 5	6	3
9	23+ 8	5	3	7	1	4	4– 2	6

327

12+ 7	10+ 6	4	1− 3	2	4− 5	16+ 1	1− 9	8
5	10+ 2	13+ 7	1− 4	3	9	6	8	1
9 9	1	6	16+ 2	12+ 4	3	20+ 8	5	7
10+ 1	3	8	6	5	19+ 4	7	15+ 2	20+ 9
2	4	3 3	10+ 9	1	8	15+ 5	7	6
4	22+ 9	8+ 2	1	21+ 8	7	3	6	5
3	8	8− 9	5	7	6	7+ 4	1	2
2− 8	5	1	1− 7	6	1− 2	15+ 9	4	10+ 3
6	2− 7	5	17+ 8	9	1	2	3	4

328

15+ 4	2	9	2− 3	5	41+ 6	7	8	1
8− 9	1	2 2	2− 6	12+ 4	8	11+ 3	5	7
5− 2	7	7+ 6	4	8− 1	9	5 5	3	8
14+ 5	8	1	16+ 7	9+ 2	3	25+ 9	4− 6	4
1	15+ 3	19+ 8	9	7 7	4	6	2	2− 5
8	4	7	2 2	11+ 6	5	1	9	3
3− 3	17+ 5	4	8− 1	9	10+ 2	8	7 7	17+ 6
6	9	3	21+ 5	8	11+ 7	2− 4	5+ 1	2
7 7	1− 6	5	8	3	1	2	4	9

329

6− 8	2	1− 5	6	6+ 3	12+ 4	8− 1	9	21+ 7
8− 9	1	19+ 3	7	2	8	15+ 4	5	6
3+ 2	18+ 7	9	6+ 4	1	18+ 3	5 5	6	8
1	3	8	2	34+ 5	6	9	11+ 7	4
17+ 4	8− 9	1	8	6	7	10+ 3	2	5
3	4	6	8− 9	8	2− 5	7	13+ 1	5+ 2
1− 5	6	13+ 2	1	3− 7	12+ 9	8	4	3
21+ 6	5 5	7	17+ 3	4	1	2	5− 8	8− 9
7	8	4	5	9	8+ 2	6	3	1

330

13+ 9	7 7	13+ 8	5	10+ 3	1	2	4	9+ 6
4	13+ 2	18+ 5	5− 8	7+ 1	2− 7	30+ 9	11+ 6	3
7	4	2	3	6	9	8	5	18+ 1
7− 8	8+ 3	4	11+ 2	5	3− 6	7	3+ 1	9
1	5	7	4	24+ 9	3	6	2	8
4− 5	9	3− 3	6	7	2	2− 1	4− 8	4
1− 2	1	6	20+ 7	8	4	3	4− 9	5
14+ 6	8	8− 9	1	2− 2	13+ 5	1− 4	4− 3	7
3− 3	6	1	9 9	4	8	5	9+ 7	2

331

17+ 6	9	2	4− 8	4+ 1	1− 7	12+ 5	4	3
20+ 7	10+ 2	8	4	3	6	4− 9	5	13+ 1
9	11+ 8	3	11+ 6	4	1	5− 7	2	5
4	15+ 5	6	8− 1	9	1− 8	8+ 2	3	7
8+ 1	7	4	5 5	3− 2	9	3	2− 8	6
9+ 3	6	12+ 1	2	5	4− 4	8	2− 7	9
9+ 5	4	21+ 7	9	15+ 8	26+ 3	11+ 1	4− 6	2
6− 8	4+ 1	5	3	7	2	6	8− 9	4− 4
2	3	9	7	6	5	4	1	8

332

18+ 9	7	10+ 3	4	3− 8	5	9+ 6	1	2
2	11+ 5	2− 4	3	30+ 6	9	7	8	6− 1
12+ 8	6	2	15+ 1	8− 9	12+ 4	13+ 3	5	7
4	18+ 3	6	8	1	2	5	2− 7	9
7	8	8− 1	14+ 2	4− 3	6	5− 9	1− 4	5
8+ 3	19+ 2	9	5	7	4+ 1	4	2− 6	8
5	9	8	7	4 4	3	5+ 1	2	13+ 6
5− 6	21+ 1	7	9	3− 5	1− 8	2	6− 3	4
1	4	1− 5	6	2	7	8 8	9	3

333

3− 6	9	15+ 4	8	3	6+ 1	11+ 7	1− 2	5 5
7+ 3	4− 7	2− 9	3+ 2	6− 8	5	4	1	14+ 6
4	3	7	1	2	3− 9	6	17+ 5	8
11+ 8	4 4	9+ 2	7	23+ 5	6	9	3	8+ 1
1	2	10+ 6	4	2− 9	7	5	8 8	3
8+ 2	6	22+ 8	5	7	17+ 3	8− 1	9	4
2− 7	11+ 1	5	9	6	8	17+ 3	2− 4	2
9	5	10+ 1	3	4	2	8	6	2− 7
16+ 5	8	3	11+ 6	1	4	9+ 2	7	9

334

16+ 8	6+ 1	5	8+ 2	3	1− 6	11+ 4	7	20+ 9
6	1− 8	7	1	2	5	5− 9	4	3
2	5− 6	1	16+ 7	5	4	12+ 3	9	8
8+ 5	6− 2	8	8− 9	1	15+ 3	13+ 7	6	5+ 4
3	16+ 7	20+ 9	6	4	2	21+ 5	12+ 8	1
16+ 9	4	6	5	7	8	1	3	8+ 2
7	5	7+ 3	4	19+ 8	9	2	1	6
15+ 4	9	2	19+ 3	6	1	20+ 8	5	7
4+ 1	3	12+ 4	8	9	1− 7	6	7+ 2	5

335

7− 1	8	17+ 5	9+ 2	4	24+ 3	7	15+ 9	6
13+ 7	6	8	4	3	9	5	10+ 1	2
13+ 9	21+ 5	3	1	2	4	6	7	11+ 8
4	3+ 1	2	2− 8	16+ 7	8+ 6	14+ 9	5	3
14+ 5	11+ 3	7	6	9	2	13+ 1	8	16+ 4
2	7	1	8+ 5	1− 6	15+ 8	4	3	9
23+ 8	9	6 6	3	5	7	6+ 2	10+ 4	1
6	13+ 4	16+ 9	7	8 8	1	3	23+ 2	5
3	2	4	15+ 9	1	5	8	6	7

336

2 2	7560× 7	6	288× 8	9	21× 3	1	40× 5	4
5	6	54× 3	5760× 9	4	7	56× 8	2	45360× 1
6	2	9	4	8	5	7	1	3
288× 9	14× 1	7	6 6	84× 2	4	120× 3	8	5
8	4	1	7	6	90× 2	5	3 3	9
63× 7	3	2	1	5	9	64× 4	6	8
3	36× 9	4	105× 5	1	8	2	7	12× 6
20× 4	5	8 8	3	7	8640× 1	6	324× 9	2
1	8	5	2	3	6	9	4	7 7

337

2160× 4	2	5	42× 7	6	1	5184× 9	8	3
30× 5	735× 7	3	12× 4	1	72× 8	6	432× 9	16× 2
6	5	2	3	56× 7	9	4	1	8
7	3	9	1	8	2400× 5	2	6	4
864× 2	6	8	5	4	3	1	441× 7	9
9	96× 8	6	2	45× 5	84× 7	3	4	1
8	12× 4	1	864× 6	9	2	2250× 5	3	7
3	1	252× 7	9	2	4	112× 8	5	6
1	9	4	144× 8	3	6	7	2	5

338

60× 1	5	144× 3	8	6804× 2	9	168× 4	7	6
135× 5	1	2	6	20× 4	3	7	9	56× 8
9	3	6	1	5	32× 4	8	2	7
1470× 3	7	5	12× 2	6	25920× 8	1	288× 4	9
2	48× 4	1	7	168× 3	6	9	8	5
6	2	1260× 9	5	8	7	3	1	4
32× 8	432× 6	7	4	63× 9	1	150× 5	6× 3	2
4	9	8	27× 3	7	20× 2	6	5	1
224× 7	8	4	9	1	5	2	18× 6	3

339

960× 5	8	4	1470× 3	7	2	9 9	162× 6	1
6	96× 4	2016× 8	2× 2	1	7	5	9	3
8	3	9	7	960× 4	84× 6	2	10× 1	5 5
4860× 9	6	3	4	8	1	7	5	2
126× 7	9	1	24× 8	3	5	6	8× 2	4
2	5	6	9× 1	9	768× 8	4	126× 3	7
56× 1	7 7	10× 2	29160× 9	5	4	3	8	6
4	1	5	6	60× 2	3	56× 8	7	288× 9
3 3	2	7	5	6	9	1	4	8

340

135× 9	252× 4	168× 7	3	8	8400× 6	5	108× 2	1
3	7	1	16× 8	2	5	9	6	4536× 4
5	9	80× 2	12× 4	3	8	7	48× 1	6
4032× 8	3	5	1	72× 9	2	6	4	7
7	6	8	45× 5	4	378× 1	2	3	9
4	160× 2	18× 3	9	6	7	1	8 8	75× 5
2	8	6	42× 7	1	9	672× 4	5	3
216× 1	5	4	6	7	3	8	189× 9	16× 2
6	1	9	40× 2	5	4	3	7	8

341

40× 5	2	126× 7	48× 6	1	216× 8	3	180× 9	4
12× 6	4	2	8	504× 3	7	9	5	1
2	1	9	4	6	40× 5	8	504× 7	3
9 9	15× 5	3	14× 7	2	1	40× 4	6	56× 8
56× 8	144× 3	1	108× 2	9	6	5	4	7
7	6	4 4	225× 1	105× 5	3	2	24× 8	54× 9
12× 4	8	5	9	7	72× 2	1	3	6
3	504× 7	8	5	4	9	42× 6	1	20× 2
1	9	576× 6	3	8	4	7	2	5

342

432× 8	1	9	2268× 6	84× 4	3	7	40× 5	2
3	480× 5	2	7	6	9	64× 8	1	4
5 5	8	3024× 6	9	1	4	2	252× 3	56× 7
4	3	1	6480× 5	9	7	6	2	8
18× 9	2	7	8	3	75× 5	1	24× 4	6
14× 2	4320× 9	5	4	8	6	3	336× 7	1
7	4	12× 3	1	84× 2	8 8	5	6	43740× 9
42× 1	6	4	3	7	2	9	8	5
6	7	80× 8	2	5	1	4	9	3

343

4 4	3 5	280 7	8	8 1	9	17 2	16 6	3
1	2	5	224 4	8	6	9	60 3	7
243 3	9	24 4	7	2880 6	2	8	5	1
9	32 8	2	3	21 7	1	5	4	22 6
280 8	4	1	108 2	5 5	3	6	7	9
5	336 7	8	6	9	10 4	3 3	1	2 2
7	54 3	6	8 9	64 2	5	1	2 8	4
9 2	6	3	1	4	8	3136 7	45 9	5
6	1	135 9	5	3	7	4	2	8

344

42 7	6	4 1	4	17 8	9	9072 3	20 5	2
3 1	2	4 3	19 5	6	8	7	4	9
160 5	4	7	6 3	2	12 1	8	9	6
8	8 9	5 2	7	1	5	6	126 3	12 4
14 2	1	2 6	280 8	5	180 4	9	7	3
9	3	4	1 1	7	3 2	5	6	15 8
5400 3	5	8	11 2	9	6	64 4	1	7
224 4	8	9	216 6	21 3	7	6 1	2	4 5
6 6	7	5	9	4	3	2	8	1

345

(15) 2	8	(56) 4	(120) 5	6	(8) 1	(10) 7	(2) 3	(17) 9
5	2	1	7	4	9	3	6	8
(17) 8	9	(168) 7	3	(7) 1	2	4	(150) 5	6
(4) 1	4	2	(9072) 6	7	3	(17) 8	9	5
4	1	(162) 3	9	(300) 5	6	(18) 2	(56) 8	7
(21) 3	7	6	8	2	5	9	(3) 4	1
(30) 6	5	9	(4) 4	(1) 8	7	(3) 1	2	(252) 3
(22) 9	6	(18) 8	1	(3) 3	(2) 4	(5) 5	7	2
7	3	5	2	9	8	6	1	(4) 4

346

(252) 9	1	7	4	(3) 6	2	(12) 3	(25) 8	(2) 5
(3) 1	(24) 2	(1920) 5	6	8	(63) 9	4	3	7
2	3	4	8	(4) 1	7	(432) 6	5	9
(1960) 7	(12) 6	2	1	5	8	9	(4608) 4	3
5	(17) 8	9	(126) 7	3	6	(3) 2	1	4
8	7	(162) 6	9	(2) 4	(30) 3	5	2	1
(13) 6	(180) 9	3	(8) 5	2	(4) 4	(294) 1	7	8
4	5	(7) 8	3	(8) 9	1	7	6	2
3	4	1	(70) 2	7	5	(17) 8	9	6

347

(1) 4	5	(384) 6	8	(3) 2	1	(567) 9	3	7
(336) 7	(252) 1	8	(9) 4	5	(108) 9	6	2	3
8	4	7	3	(42) 6	(3) 2	1	(4) 5	9
6	(17) 9	3	1	7	(4) 8	2	(80) 4	5
(6) 1	8	(2) 2	(270) 5	9	6	(168) 3	7	4
3	2	(15) 9	6	(12) 4	(1400) 7	5	8	(12) 1
(5) 2	7	(36) 4	9	3	5	8	(8) 1	6
(135) 5	(42) 6	1	7	(11) 8	3	(15) 4	9	2
9	3	(8) 5	2	1	4	7	(2) 6	8

348

(192) 8	6	2	(21) 1	3	(6) 5	(2) 7	9	(25) 4
2	(17) 8	9	7	(80) 4	1	(90) 3	5	6
(8) 9	(3) 1	(1) 3	2	5	4	6	7	8
1	2	(15) 4	3	(19) 9	(26) 7	8	6	5
(80) 5	4	8	(48) 6	7	3	(3) 2	1	(189) 9
4	(45) 9	1	8	(12) 2	(30) 6	5	3	7
(105) 7	3	5	(4) 4	6	(23) 8	9	(6) 2	1
(2) 6	5	(63) 7	9	1	2	4	(64) 8	3
3	(42) 7	6	(13) 5	8	(8) 9	1	4	2

349

(2) 3	6	(30) 5	2	(256) 4	8	(17) 1	9	7
(2) 2	(7) 9	3	(15) 6	8	(140) 5	7	4	(8) 1
(2) 8	2	6	3	(18) 5	7	(17) 9	1	4
4	(4) 5	9	(3) 1	3	6	8	(112) 7	2
(189) 9	3	7	(3) 4	(36) 6	1	(300) 5	2	8
(210) 5	(7) 8	(2) 1	7	2	3	4	(324) 6	9
7	1	2	(17) 8	9	(21) 4	3	5	6
6	(140) 7	4	5	1	9	(6) 2	8	(11) 3
(3) 1	4	(17) 8	9	7	(3) 2	6	3	5

350

(2) 4	(3) 6	(22) 9	(24) 8	3	1	(14) 2	(9) 5	(14) 7
6	3	5	(441) 9	1	7	8	4	2
(2) 2	(17) 9	8	7	(4) 6	(60) 5	4	(5) 3	1
(17) 9	8	(1008) 7	6	2	4	3	1	(2) 5
8	4	6	(1) 1	(16) 7	9	(10) 5	2	3
(105) 3	(5) 7	2	(240) 5	8	6	(8) 1	9	(288) 4
7	(3) 2	1	(960) 4	5	3	(15) 9	6	8
5	(5) 1	(7) 3	2	(4) 4	8	(15) 6	(15) 7	9
1	5	4	(3) 3	9	2	7	8	(6) 6

351

(105) 5	3	(1) 9	8	(14) 1	7	(18) 6	(2) 2	4
7	(9) 9	(70) 2	(7) 3	6	8	4	(20) 1	5
(2) 2	5	7	1	3	(216) 9	8	4	(1) 6
4	(32) 8	1	(5) 5	(567) 9	3	(36) 2	6	7
(48) 8	6	4	9	7	(25) 5	(8) 1	3	(42) 2
(252) 6	2	(192) 8	4	5	1	9	7	3
3	7	6	(48) 2	(2) 8	4	(5) 5	(8) 9	1
(8) 9	(8) 1	3	6	4	(2) 2	(105) 7	5	(25) 8
1	4	(35) 5	7	(3) 2	6	3	8	9

352

(150) 6	5	(7) 4	2	1	(24) 3	(504) 7	9	8
5	(3) 3	(8) 9	(48) 6	(22) 8	4	2	(126) 7	(20) 1
(28) 7	(17) 2	1	8	5	9	3	6	4
4	9	6	(12) 3	7	2	(40) 1	8	5
1	(15) 8	2	(4) 4	(216) 9	6	5	(10) 3	7
(4) 3	7	5	1	4	(8) 8	(378) 9	(4) 2	6
(19) 8	(15) 1	3	(90) 5	2	7	6	(160) 4	(9) 9
2	4	7	9	(2) 6	(6) 1	8	5	(6) 3
9	(336) 6	8	7	3	5	(4) 4	1	2

353

(140) 7	5	(8) 9	(7) 4	(4) 8	2	(1) 6	(6) 3	(16) 1
4	(6) 6	1	2	(96) 3	8	5	9	7
(45) 9	(3) 2	6	1	4	(1) 5	(147) 3	7	8
5	(14) 3	(4) 4	(8) 9	1	6	7	(13) 8	(12) 2
3	1	(320) 5	8	(20) 9	(5) 7	2	4	6
(3) 2	7	8	6	5	(3) 3	(360) 9	1	4
6	(15) 4	3	(105) 7	(108) 2	1	8	5	(3) 9
(18) 1	8	(14) 7	5	6	9	(4) 4	(4) 2	3
8	9	2	3	(3) 7	4	1	6	(5) 5

354

(2) 4	6	(10) 1	(63) 7	9	(19) 8	(5) 5	(5) 2	3
(120) 3	8	9	(4) 5	1	7	4	(10) 6	(11) 2
(2) 7	5	(1) 2	(11) 8	(16) 6	(3) 1	3	4	9
5	(7) 7	3	1	8	2	(54) 6	(63) 9	(2) 4
(8) 1	(2) 4	(40) 8	2	(3) 3	(135) 5	9	7	6
9	2	5	(12) 6	4	3	(7) 8	(1) 1	(15) 7
(3) 6	(3) 3	(3) 7	4	2	9	1	(14) 5	8
2	9	(162) 6	3	(20) 5	4	(14) 7	8	1
(7) 8	1	(4) 4	9	(42) 7	6	2	(2) 3	5

355

(4) 7	(14) 5	(17) 9	(96) 2	6	8	(8) 4	1	3
3	1	8	(210) 5	(10) 4	(3) 9	6	(20) 2	7
(20) 5	8	6	7	1	(12) 4	3	9	2
4	(18) 9	(12) 3	(17) 6	2	(35) 5	(19) 8	7	(216) 1
(6) 6	2	5	8	3	7	1	4	9
(12) 8	7	4	3	(360) 9	1	(3) 2	5	6
1	3	(63) 7	9	8	(3) 2	5	(2) 6	4
(7) 9	(2) 4	(1) 2	1	5	(1) 6	7	3	(2880) 8
2	6	1	(14) 4	7	3	9	8	5

356

(1280) 8	(4) 7	3	(8) 1	(3) 6	(45) 9	5	(2) 4	2
4	5	(15) 6	7	2	(14) 3	(72) 9	8	(4) 1
(567) 9	8	7	2	(3360) 3	4	6	1	5
7	9	(576) 8	4	5	(9) 6	(2) 1	(1) 2	3
(1) 3	4	(8) 9	6	8	1	2	(2) 5	7
(17) 5	6	1	3	4	2	(7) 7	(1) 9	8
6	(2) 1	(72) 2	9	7	(960) 5	8	3	(2) 4
1	2	4	(3) 5	(8) 9	8	(4) 3	7	6
(2) 2	(2) 3	5	8	1	(28) 7	4	(15) 6	9

357

(2) 1	2	(1) 3	(2) 4	(48) 6	8	(63) 9	(2) 5	7
(13) 4	9	2	6	(36) 3	7	1	(3) 8	5
(6) 7	1	(16) 8	2	4	3	(1) 5	6	(26) 9
(1) 2	3	1	(15) 5	7	(960) 4	6	9	8
(1) 9	(19) 6	4	3	(2) 2	5	8	(84) 7	1
8	(200) 5	9	(17) 7	(10) 1	2	3	4	6
5	8	(35) 7	1	9	(6) 6	(3) 4	(5) 3	2
(2) 6	(11) 4	5	(23) 9	(7) 8	1	7	2	(8) 3
3	7	6	8	(16) 5	9	2	1	4

358

(30) 2	3	5	(17) 7	6	4	(90) 9	(7) 8	(27) 1
(720) 4	5	(30) 6	9	7	8	2	1	3
(48) 6	4	(28) 8	(6) 1	(9) 2	3	5	(20) 7	9
8	9	3	5	4	(2) 2	1	6	7
(7) 7	(3) 6	9	(7) 4	3	(7) 1	8	(3) 2	5
(10) 1	2	7	(2) 6	(270) 9	(2) 5	3	(15) 4	8
9	(280) 7	1	8	5	6	(4) 4	3	(12) 2
5	8	(42) 2	3	1	7	(1) 6	(14) 9	4
(10) 3	1	4	2	(1) 8	9	7	5	6

359

(14) 2	7	(54) 6	(3) 1	3	(1) 4	5	(24) 9	8
(2) 4	1	9	(30) 6	5	(240) 8	(5) 2	3	7
8	(10) 2	3	4	1	5	6	(17) 7	(36) 9
(9) 1	3	5	(2) 2	(17) 8	9	(147) 7	6	4
(162) 6	(160) 8	(10) 2	5	(9) 9	7	3	4	1
3	4	(1) 8	(5) 7	2	(1) 1	(20) 9	5	6
9	5	7	(3) 3	(192) 4	6	8	1	(1) 2
(21) 5	(5) 6	1	9	(5) 7	2	(32) 4	8	3
7	9	(2) 4	8	(3) 6	3	1	(7) 2	5

360

(90) 3	6	(10) 2	(4) 4	1	(19) 7	(6) 8	(180) 5	9
5	7	1	(14) 6	(8) 3	8	2	(17) 9	4
(15) 6	2	7	1	5	4	(9) 9	8	(5) 3
7	(81) 3	9	(5) 5	(48) 4	2	6	(10) 1	8
(3) 1	4	3	(4) 2	(360) 8	9	5	7	(6) 6
(7) 4	(3) 9	6	8	(105) 7	5	3	2	(4) 1
2	1	(24) 8	7	(18) 9	(2) 6	4	(2) 3	5
(576) 8	(100) 5	4	9	2	(2) 3	(6) 1	6	(56) 7
9	8	5	(3) 3	6	1	7	4	2

361

(4) 9	(42) 7	(3) 2	6	(3) 1	(4) 8	4	(10) 3	(12) 5
5	6	(5) 3	(162) 9	4	(13) 1	8	2	7
(63) 7	9	8	3	6	4	(10) 5	1	(2) 2
(2) 6	3	(1) 7	8	(14) 5	9	2	4	1
(10) 2	5	(24) 4	(21) 7	3	(9) 6	1	(8) 8	(13) 9
3	(128) 8	6	1	(126) 7	2	(45) 9	5	4
8	2	1	(1) 4	9	(2) 5	(7) 7	(2) 6	3
(6) 1	(13) 4	9	5	2	3	(1008) 6	7	8
4	1	(22) 5	2	8	7	3	(54) 9	6

362

(2160) 9	4	6	(63) 1	(20) 8	3	(14) 2	5	7
5	(240) 6	8	7	3	9	(7) 1	2	4
2	5	(4) 4	3	(7) 1	8	(84) 7	(48) 6	(27) 9
(126) 7	9	1	(160) 4	5	2	6	8	3
(4) 4	2	(240) 5	8	(22) 6	7	(6) 3	9	1
1	8	3	(4) 5	9	(11) 6	4	(672) 7	(2) 2
(144) 6	(11) 7	2	9	(3) 4	1	(1800) 5	3	8
3	1	(7) 9	2	7	5	8	4	(12) 6
8	3	(336) 7	6	2	4	9	1	5

363

2	1	3	6
3	2	6	1
6	3	1	2
1	6	2	3

Cage clues: 18, 6, 3, 8, 9, 1, 5

364

4	6	7	8	2
2	7	8	4	6
8	4	6	2	7
7	8	2	6	4
6	2	4	7	8

Cage clues: 2688, 4, 19, 17, 2, 24, 1, 6, 56, 8

365

9	4	8	1	3	6
8	3	4	9	6	1
1	6	3	4	9	8
3	8	9	6	1	4
6	9	1	8	4	3
4	1	6	3	8	9

Cage clues: 1152, 20, 5, 6, 5, 27, 18, 1, 2, 8, 32, 1, 14, 2, 9

Printed in the United States
By Bookmasters